Researching Patient Safety and Quality in Healthcare

A NORDIC PERSPECTIVE

Researching Patient Safety and Quality in Healthcare

A NORDIC PERSPECTIVE

EDITED BY KARINA AASE
LENE SCHIBEVAAG

CRC Press is an imprint of the
Taylor & Francis Group, an **informa** business

CRC Press
Taylor & Francis Group
6000 Broken Sound Parkway NW, Suite 300
Boca Raton, FL 33487-2742

First issued in paperback 2019

ISBN-13: 978-1-4724-7713-2 (hbk)
ISBN-13: 978-0-367-88173-3 (pbk)

Library of Congress Cataloging-in-Publication Data

Names: Aase, Karina, editor. | Schibevaag, Lene, editor.
Title: Researching patient safety and quality in healthcare : a Nordic perspective / Karina Aase and Lene Schibevaag.
Description: Boca Raton : Taylor & Francis, 2017.
Identifiers: LCCN 2016011478 | ISBN 9781472477132 (hard cover)
Subjects: LCSH: Medical care--Norway--Quality control. | Medical care--Sweden--Quality control. | Medical care--Denmark--Quality control. | Patients--Norway--Safety measures. | Patients--Sweden--Safety measures. | Patients--Denmark--Safety measures.
Classification: LCC RA399.N8 R43 2017 | DDC 362.109481--dc23
LC record available at http://lccn.loc.gov/2016011478

Visit the Taylor & Francis Web site at
http://www.taylorandfrancis.com

and the CRC Press Web site at
http://www.crcpress.com

Contents

Editors

Professor Karina Aase, University of Stavanger, is an internationally recognised quality and safety expert with a multi-sector background. She is the founder of the *Quality and Safety in Healthcare Systems* research group at the Department of Health. Under Aase's leadership, this group has become Norway's largest and most influential multidisciplinary research group within the field of healthcare quality and patient safety, boasting more than 20 affiliated researchers, as well as numerous national and international collaborators. Professor Aase has a proven record of gaining and leading large funding grants. Aase has a publication record in merited journals, has edited several books on patient safety and quality in healthcare and is an associate editor of BMC Health Services Research. She is a member of the National Council for the Patient Safety Programme and has currently been appointed by the King in Council as a member of a government committee proposed to suggest a new organisation of state ownership of the specialist healthcare services.

Lene Schibevaag has a bachelor degree in physiotherapy from the University of Northumbria and has worked for several years as a physiotherapist in both a hospital and a municipality setting in England and Norway. She has a master's degree in societal safety from the University of Stavanger with a dissertation on transitional care entitled 'Multidisciplinary Collaboration: A study of the physiotherapist role in transitional care of the elderly'. Schibevaag is currently working as a research coordinator for the 'Quality and Safety in Healthcare' research group at the University of Stavanger, Norway.

Contributors

Marja Äijö, PhD, is a principal lecturer at Savonia University of Applied Sciences, School of Health Care, Finland. Äijös' teaching field is gerontology and her main research interests are physical activity, cardiovascular diseases and all-cause mortality among older people and simulation teaching and learning among healthcare students.

Henning Boje Andersen is a professor of healthcare management at the Technical University of Denmark (DTU) in the Department of Management Engineering. His background is in philosophy and human factors, and his R&D activities, beginning at Risø National Laboratory in 1984 and continued at DTU since 2008, have been focused on human and organisational factors that influence performance in safety critical domains including aviation, maritime operations and healthcare. His recent research and teaching projects include the design and implementation of healthcare technology for quality improvement, tele-health and independent living for the elderly.

Roland Bal is a professor of healthcare governance at Erasmus University in Rotterdam, the Netherlands. He obtained his MSc in health science and his PhD in science and technology studies. His research interests include the relation between science, policy and practice and governance infrastructures in healthcare. He led the evaluation of several large-scale evaluations of quality and safety of care programs in the Netherlands and participated in the EU-funded QUASER project on quality management in European hospitals. Related to his work on quality, he participates in research on information technologies in healthcare and the ways in which healthcare organisations 'organise for transparency'. He is a co-founder of the academic collaborative on supervision in healthcare, which is run by the Healthcare Inspectorate of the Netherlands.

Paul D. Bartels is an MD and a specialist in chemical pathology. He is the director of the Danish Clinical Registries – a national quality improvement programme. Also, he is the head of the European Society for Quality in Healthcare (ESQH) Office for Quality Indicators, Aarhus, and an ESQH fellow (past president of ESQH). Bartels is a former medical director of Randers Central Hospital and a former senior lecturer in chemical pathology, University of Aarhus, and a former chairman of the Danish Committee for Quality Assurance in Chemical Pathology. He is a clinical lecturer at Aalborg University. Bartels is the author and co-author of numerous publications on quality and patient safety in healthcare.

Inger Johanne Bergerød is a PhD candidate in health and medicine at the University of Stavanger. She has worked as an oncology nurse and middle management in the Department of Hematology and Oncology at Stavanger University since 1999. Bergerød holds a master's degree in health science from 2012 (thesis title: 'Leadership, quality and patient safety – A comparative case study of two Norwegian hospitals'). The research for the master thesis was done in collaboration with the University of Stavanger in the EU FP7 project 'Quality and Safety in European Hospitals'. The title of Bergerød's commenced PhD study is 'Improving the quality and safety of cancer care: A study of next-of-kin involvement'. Key research interests are oncology, nursing, leadership, patient safety, quality improvement, next-of-kin, organisational safety and cultural factors.

Hester van de Bovenkamp, PhD, is an associate professor of public administration in healthcare at Erasmus University Rotterdam, the Netherlands. Her main field of interest is citizen participation and representation in healthcare decision-making. Her research is focused on the various forms of citizen participation and representation that can be found at different levels of the healthcare system (micro, meso, macro). She participated in numerous studies on this subject. She also studies the governance of healthcare quality using an institutional perspective. On this subject she participated in several studies including the EU-funded QUASER project which focused on quality work in hospitals across Europe.

Jeffrey Braithwaite, PhD, is the foundation director of the Australian Institute of Health Innovation, a director of the Centre for Healthcare Resilience and Implementation Science and a professor of Health Systems Research, Faculty of Medicine and Health Sciences, Macquarie University, Australia. His research examines the changing nature of health systems, particularly patient safety, standards and accreditation, leadership and management, the structure and culture of organisations and their network characteristics, which has attracted funding of some AUD$90 million. He holds visiting professorial appointments at the University of Birmingham, UK; Newcastle University, UK; the University of Southern Denmark; the University of New South Wales, Australia and the Canon Institute of Global Studies, Tokyo, Japan.

Angela Coulter is a health policy analyst and a researcher who specialises in patient and public involvement in healthcare. She is the director of Global Initiatives at the Informed Medical Decisions Foundation, Boston, and a senior research scientist in the Nuffield Department of Population Health, University of Oxford. A social scientist by training, Angela, has a doctorate in health services research from the University of London. From 2000 to 2008, she was the chief executive of Picker Institute Europe. Previous roles included director of Policy and Development at the King's Fund and director of the Health Services Research Unit at the University of Oxford. She is an adjunct professor at the University of Southern Denmark and a senior visiting fellow at the King's Fund in London, holds honorary fellowships at the UK Faculty of Public Health and the Royal College of General Practitioners and is a trustee of National Voices.

Angela has published more than 300 research papers and reports and several books including *The Autonomous Patient*, *The European Patient of the Future* (winner of the 2004 Baxter Award), *The Global Challenge of Healthcare Rationing*, *Hospital Referrals*, *Engaging Patients in Healthcare* (highly commended by the BMA) and *Understanding and Using Health Experiences*. She was the founding editor of *Health Expectations*, an international peer-reviewed journal on patient and public involvement in healthcare and health policy. She has won awards for her work from the Donabedian Foundation of Barcelona in 2012 and the International Shared Decision Making Conference in 2013.

Mirjam Ekstedt, RN, is an associate professor at the Royal Institute of Technology and at the Medical Management Centre, LIME, at Karolinska Institutet in Stockholm. She is leading a research program on systems safety in healthcare, with focus on management and decision-making at the sharp end of complex organisations, patients' and family caregivers' involvement and implementation of patient-centred innovations and communication systems. She gained her PhD in psychosocial medicine at Karolinska Institutet with focus on burnout and sleep and effects of fatigue and stress on health and performance.

Carsten Engel, MD, is the deputy chief executive at IKAS, the Danish Institute for Quality and Accreditation in Healthcare. His background is clinical practice as anaesthesiologist and management experience at the departmental and hospital level. Since 2004, he has devoted full time to quality management and improvement in healthcare, taking a leading part in the development and management of the Danish healthcare accreditation programme (DDKM). Through the Accreditation Council of the International Society of Quality in Health Care (ISQua), he is engaged in accreditation internationally and he serves as an ISQua Expert. While not primarily a researcher, he takes a keen interest in supporting research in accreditation and quality improvement strategies in general and in spreading results of such research.

Veslemøy Guise, MA, is a PhD candidate at the Department of Health Studies, University of Stavanger, Norway. Guise is a medical sociologist by background and has worked as a healthcare researcher in New Zealand and the United Kingdom in addition to Norway. Her current research is centred on the *Safer@Home – Simulation and training* and is concerned with the development, implementation and evaluation of telecare training programs for healthcare professionals working in municipal home healthcare services.

Kaisa Haatainen, PhD, is an adjunct professor (mental health promotion) of nursing sciences at the University of Eastern Finland. For the last three years, she has been working as a patient safety manager at Kuopio University Hospital. Coordinating patient safety at the hospital level and developing the reporting and measuring of it are in the focus of her daily job. Dr. Haatainen has a membership as an expert at the Finnish Centre for Evidence-Based Health Care, an affiliated centre of the Joanna Briggs Institute. Her research interests are in mental well-being and patient safety. Currently, she is leading a research project focusing on patient safety.

Britt Sætre Hansen is a professor in nursing and an intensive care nurse with a master's degree from the University of Bergen and PhD from the University of Stavanger in 2009. Her PhD thesis is about differences in intensive care nurses' and physicians' inter- and intraprofessional understanding of protocol-directed weaning from mechanical ventilation. She is employed at the University of Stavanger and Buskerud and Vestfold University College where she is supervising and teaching master students in research methods. Her research relates to interprofessional collaboration and implementation of evidence-based knowledge. She is also employed as a senior advisor at Stavanger University Hospital and works with research, professional development and dissemination of research in clinical practice.

Arja Holopainen, PhD, is a research director at the Nursing Research Foundation. Holopainen is also the director of two international collaborating centres in Finland: (1) the Finnish Centre of Evidence-Based Health Care (Joanna Briggs Institute) and (2) the WHO Collaborating Centre for Nursing. Her main research interest concerns evidence-based health care and how to synthesis evidence (systematic reviews), as well as how to improve and to evaluate evidence-based practices.

Sissel Eikeland Husebø is a postdoctor candidate and has experience within a wide variety of acute care clinical settings. She is an intensive care nurse and has worked at SAFER simulation centre (www.safer.net) for seven years designing, implementing and facilitating simulation-based learning in the nursing and healthcare. Her PhD thesis is about condition for learning in simulation with focus on team-based resuscitation in nursing education. She is currently employed at the University of Stavanger and Stavanger University Hospital where she supervises nursing, master and PhD students. Her ongoing research is on interprofessional clinical leadership in teams in the emergency department and interprofessional decision-making in the hospital setting.

Anne Marie Lunde Husebø, RN, GN, MSc, is an associate professor at the Department of Health Studies, University of Stavanger, Norway. Husebø conducts research within the field of health promotion in chronic disease and on telecare in primary care services. She defended her thesis in December 2015 titled 'Exercise during breast cancer treatment. A study of physical and psychosocial outcomes, and motivational challenges'. Husebø has vast experience as a nurse specialist (geriatric nursing) working in long-term care facilities for the elderly.

Søren Paaske Johnsen is a research consultant and an associate professor in clinical epidemiology at Aarhus University Hospital and Aarhus University, Denmark. Dr. Johnsen's primary field of interest is quality of clinical care, including the use of evidence-based care, effectiveness and safety of recommended clinical interventions and evaluation of quality improvement strategies. The activities are primarily focused on the scientific use of clinical quality databases and administrative registers. Dr. Johnsen received his medical degree and PhD degree in epidemiology from Aarhus University. He has worked with clinical quality databases since the late 1990s and for 10 years was the head of Centre for Clinical Databases at the Department of Clinical Epidemiology, Aarhus University Hospital. This centre is one of three national Danish centres supporting clinical databases with expertise in clinical epidemiology and biostatistics. He currently leads a research group working with quality of care at Aarhus University Hospital and Aarhus University.

Eija Kivekäs, MSc (health management), RN, is a project researcher and a doctoral student in Health and Human Services Informatics at the University of Eastern Finland, Department of Health and Social Management. Kivekäs has coordinated several clinical development projects and worked as a project manager for several years. Her research interest is the role of patient safety in electronic health records as well as competencies and management issues in information system implementation.

Hannu Kokki, MD, PhD, is a professor of anaesthesiology at the University of Eastern Finland, School of Medicine, at Kuopio Campus. He is the chair of the committee on pharmacotherapy at the Hospital District of Northern Savo in Kuopio University Hospital and a clinical expert for the Finnish Office for Health Technology Assessment and Patients Insurance Centre. Hannu Kokki has a large research network in clinical medicine, pharmacology, social pharmacy and nursing sciences.

Pia Kjær Kristensen, MHSc, is a PhD student at the University of Aarhus, Department of Clinical Epidemiology, Aarhus University Hospital, and Department of Orthopaedic Surgery Horsens Regional Hospital, Denmark. Pia's primary field of interest is clinical epidemiology, including quality of clinical care, the organisation of in-hospital care and evaluation of quality improvement strategies among patients with hip fracture. Pia has conducted research within the healthcare sector since 2013 with projects covering overall associations between selected aspects of structure of care (orthogeriatric specialisation and patient volume) and process performance measures, 30-day mortality and hospital bed-day use. Currently, she is a part of the research group working with quality of care at Aarhus University Hospital and Aarhus University.

Jan Mainz is a professor of quality improvement and health services research at Aalborg University Hospital and an adjunct professor at the University of Southern Denmark. He is the deputy director of psychiatry in North Denmark Region. Jan Mainz is the Danish representative in the OECD Health Care Quality Indicator Project since 2002 and member of the OECD Bureau. He is also the member of WHO's Advisory Committee of the WHO Report on Safety and Quality of Health Care Services. He is the former project manager and he has been responsible for the development of indicators and measurements in the Danish National Indicator Project. He was the chairman of the Nordic Ministers Council Project on indicators monitoring in the six Nordic countries (Sweden, Norway, Finland, Iceland, Greenland and Denmark) in 2002–2012. From 1999 to 2005, he was the president of the Danish Society for Quality in Health Care, and from 2003 to 2005, president of the European Society for Quality in Health. His main research activities relate to quality improvement and quality management, performance and outcome measurement, quality monitoring and health services research and patient empowerment.

Tanja Manser is a full professor for patient safety and director of the Institute for Patient Safety at the University Hospital Bonn, Germany. Her research focuses on adaptive coordination in acute care settings, quality and safety of patient handoffs, clinical risk management, safety climate in healthcare and implementation of patient safety practices. She received a PhD in psychology from the University of Zurich in 2002 and completed her habilitation in work and organisational psychology at ETH Zurich in 2008.

Kirsti Lorentsen Moltu is an occupational therapist who works with the adult and elderly population in the home healthcare services in Stavanger municipality, Norway. Moltu has worked as a project coordinator in several municipality telecare projects. Her primary responsibility has been concerned with patient follow-up and training during the implementation of new technologies. She has also had an active role as the municipality's representative in the *Safer@Home – Simulation and training* research project.

Heidi Helen Nedreskår has a bachelor's degree in nursing, with specialisation in intensive and geriatric nursing, and a master's degree in health science. In connection with health studies at the University of Stavanger, she has been affiliated to the project 'Quality and safety in transitional care of the elderly'. Nedreskår has many years of clinical experience in both hospital and municipality services and has held different positions in management and administration in healthcare service. Nedreskår is now holding an administrative position in Stavanger municipality working towards palliative care, dementia care and patients in need of care in institutions.

Kaija Saranto, PhD, FACMI, FAAN, works as a full professor in health and human services informatics at the University of Eastern Finland, Department of Health and Social Management. In 2012, the HHSI master's degree programme received the status 'IMIA accredited' as the very first internationally. Dr. Saranto has a number of memberships in both academic and expert groups focusing on ICT use in the society. Dr. Saranto acts also as the deputy director at the Finnish Centre for Evidence-Based Healthcare, an affiliated centre of the Joanna Briggs Institute. Her research interest is the role of classifications in EHRs as well as educational and organisational issues in information system implementation. Currently, she is also leading a research group focusing on patient safety.

Annemiek Stoopendaal, PhD, is an assistant professor of organisational anthropology in healthcare at the Erasmus University Rotterdam, Department of Health Policy and Management. Her research interests are on healthcare management and governance, with specific emphasis on the work and behaviour of healthcare managers, complex interventions and the governance and supervision of quality and safety in healthcare. She is interested in the boundaries and bindings between different levels in the organisation of healthcare. She participates in the academic collaborative on supervision in healthcare, which is run by the Healthcare Inspectorate of the Netherlands and is involved in research on regulation and supervision. The research methods she uses are qualitative, ethnographic and formative research.

Marianne Storm is an associate professor at the Institute of Health Sciences at the University of Stavanger. She holds a PhD in management from the University of Stavanger and a master's degree in health science from the University of Bergen. She has a nursing background. In her postdoctoral studies, she was principal investigator for phase 2 of the research project 'Quality and safety in transitional care of the elderly' funded by the Norwegian Research Council (NRC). She has been an affiliated researcher in *Safer@Home*, a research project funded by the NRC involving the use of welfare technology in healthcare for the elderly. Her research areas include healthcare quality, patient experience and user involvement, transitional care, welfare technology, learning, organisational culture, intervention research and statistical analysis.

Tarja Tervo-Heikkinen, PdD, RN, is a clinical nurse consultant at the Kuopio University Hospital (KUH), Clinical Development, Education and Research Unit of Nursing (CDERUN). Dr. Tervo-Heikkinen's responsibility areas are fall prevention, pressure ulcer prevention, patient education and evidence-based nursing. She is the chairman of the RFP Network and also a member of KUH's Care Ethics Committee. Before her current post, Tervo-Heikkinen worked as a university researcher at the University of Eastern Finland and project manager at the Attractive and Health Promoting Health Care. Dr. Tervo-Heikkinen is also member of core staff of the Finnish Centre of Evidence-Based Health Care (Joanna Briggs Institute). Her doctoral thesis (2008) title was 'The Effectiveness of Nursing in Specialized Health Care'.

Siri Wiig, PhD, is a professor of quality and safety in healthcare systems at the University of Stavanger (UiS), Norway. She is part of the research centre SEROS – Centre for Risk Management and Societal Safety, UiS. Wiig has a part-time affiliation as chief engineer within organisational safety at the Petroleum Safety Authority, Norway. Wiig holds a PhD in risk management and societal safety from 2008 (title of PhD: 'Contributions to Risk Management in the Public Sector') and a master of science in societal safety and planning from 2002. Wiig has conducted research on quality and safety and risk management in three sectors: the healthcare sector (primary and specialised services), municipal sector (emergency preparedness and primary healthcare) and petroleum sector. Wiig has been full-time researcher in the EU-FP 7 project: 'Quality and Safety on European Union Hospitals' (2010–2013). Key research interests are organisational safety, human factors, risk management, societal safety, risk regulation, risk perception, patient safety, quality improvement, learning, sociotechnical systems and welfare technology.

Introduction

This book is about research on patient safety and quality in healthcare seen from a Nordic perspective. There is today an abundant literature on the topic in an international context, dominated by authors from the United States, the United Kingdom, Canada and Australia. This edited collection will try to distil the flavours of contemporary Nordic research on patient safety and quality in healthcare. In doing so, we will present the Nordic characteristics by pursuing different topics such as patient-centred care, methodology and theory, as well as showcasing a set of exemplary Nordic research contributions grouped around macro-political issues, meso-level organisational issues and micro-level clinical practice issues.

RATIONALE BEHIND THE BOOK

Due to a growing amount of research activities and research groups across the Nordic countries since the millennium, the Nordic Research Network for Safety and Quality in Healthcare (NSQH) was established in 2008. The purpose of the NSQH network (NSQH, 2016) is to promote and advance patient safety and quality research in the healthcare sector and to facilitate increased collaboration on research and utilisation of research results amongst the Nordic countries across research institutions and clinical settings. It has also been a long-term goal for the network to establish a Nordic research agenda. Following the NSQH network initiative, the first Nordic Patient Safety Conference was established in 2010 (Stockholm) succeeded by the Nordic Conference on Research in Patient Safety and Quality in Healthcare 2012 (Copenhagen) and 2014 (Stavanger).

Despite these valuable efforts to establish a more sustainable research stream in the Nordic countries, no overview of research on patient safety and quality in Nordic healthcare has so far been presented. The book will therefore meet this demand by systematically presenting research from the Nordic countries within the field. Invited scholars have reflected upon specific topics such as patient-centred care, methodology, theory and institutional layering related to the Nordic research context. Also, a set of research contributions have been selected for presentation to reflect the multi-layered nature of healthcare involving a complex set of stakeholders, service providers and activities. As such, we believe that this edited volume will create a basis for reflecting the nature of contemporary Nordic research on patient safety and quality in healthcare constituting a useful framework for crafting a research agenda and for making visible areas for knowledge sharing and cross-country research collaborations.

In an international context, the book will give an overview of the Nordic perspective on patient safety and quality research that so far has been lacking in the current literature. Identifying and describing characteristics of the Nordic perspective might thus benefit researchers and practitioners in other contexts.

WHAT IS PATIENT SAFETY AND QUALITY IN HEALTHCARE?

The conceptual pair of patient safety and quality is employed in many contexts of today's healthcare systems and also research wise. The relationship between patient safety and quality can be expressed in several ways, the most common being to view quality in healthcare as the overarching umbrella under which patient safety resides. For example, the Institute of Medicine (IOM) considers patient safety as indistinguishable from the delivery of quality healthcare. An early definition of quality in healthcare is expressed by the IOM as follows:

> The degree to which health services for individuals and populations increase the likelihood of desired health outcomes and are consistent with current professional knowledge
>
> **Kohn et al. (2000); Berg and Cassels (1992)**

After the millennium, descriptions of quality have been centred around conceptual dimensions such as the six dimensions suggested by IOM (2000):

1. *Safe* – avoiding injuries to patients from the care that is supposed to help them
2. *Effective* – providing services based on scientific knowledge to all who could benefit and refraining from providing services to those not likely to benefit (avoiding underuse and overuse)
3. *Patient centred* – providing care that is respectful of and responsive to individual patient preferences, needs and values and ensuring that patient values guide all clinical decisions
4. *Timely* – reducing waits and sometimes harmful delays for both those who receive and those who give care
5. *Efficient* – avoiding waste, in particular waste of equipment, supplies, ideas and energy
6. *Equitable* – providing care that does not vary in quality because of personal characteristics, such as gender, ethnicity, geographic location and socio-economic status

Several countries now base their definition and description of quality on all or several of the six dimensions (Wiig et al., 2014), including the Nordic countries. The patient safety dimension is most often related to the avoidance or prevention of patient harm. A well-established definition of patient safety is made by Charles Vincent:

> The avoidance, prevention and amelioration of adverse outcomes or injuries stemming from the process of healthcare
>
> **Vincent (2010)**

As such, patient safety is an attribute of healthcare systems; it minimises the incidence and impact of, and maximises recovery from, adverse events (Emmanuel et al., 2005). Adverse events could be seen as the outcome of poor patient safety but also as something inevitable in healthcare processes. The World Healthcare Organisation (WHO) defines adverse events as follows:

> An injury related to medical management, in contrast to complications of disease. Medical management includes all aspects of care, including diagnosis and treatment, failure to diagnose or treat, and the systems and equipment used to deliver care. Adverse events may be preventable or non-preventable
>
> **WHO (2005)**

To assess whether an injury is caused by clinical treatment or care and not by the course of the disease is in many cases not a straightforward task. Adverse events often involve complex chains of action where competence, accessible treatment methods, time and patient conditions are involved.

In addition to the landmark report 'To err is human' (US; IOM, 2000), a few large-scale system failures such as the Bristol Royal Infirmary Case (UK; Kennedy, 2001) and the Mid Staffordshire Case (UK; Francis, 2013) have been important drivers for prioritisation, focus and research on patient safety internationally. Nationally, there are also cases in which high-profile inquiries have been raised, for example, the Radboud Case in the Netherlands (Dutch Patient Safety Board, 2008). Paramount to the understanding of these complex safety cases is the system perspective applied to document the inter-relational or multi-layered context (e.g. managerial factors, professional factors, organisational factors, cultural factors) behind the development of the crises.

Unlike the patient safety field, quality and quality improvement have a longer research tradition within healthcare. While the patient safety field so far has been occupied mainly with establishing the degree of adverse events and the causal drivers behind them, the quality field has been driven by a more applied focus of improvement processes (see also Chapter 4). The quality research focuses more on finding and documenting effective processes gaining desirable outcomes, for example, as shown in the Michigan Intensive Care Unit project, which attracted international attention by successfully reducing rates of central venous catheter bloodstream infections (Dixon-Woods, 2011).

The amount of international research studies within patient safety and quality has increased rapidly over the last two decades. Still, much of the evidence is from developed nations (Jha et al., 2010), and while data from transitional and developing countries are increasing, more studies are needed to cover the global nature of patient safety and quality in healthcare. Major stakeholders behind the international patient safety and quality agenda are amongst others the WHO, Institute for Healthcare Improvement and Agency for Healthcare Research and Quality. Some would claim that the focus so far has been dominated by an instrumental approach focusing on formal and structural dimensions to develop evidence-based tools, measurements and resources (Pronovost et al., 2015).

Research principles to guide future contributions are therefore suggested as follows (Iedema, 2009):

1. Research must engage with both the predictability and the complexity of the sites and processes it seeks to describe, explain and/or impact on.
2. Engaging with complexity implicates researchers in experiencing it in a process of sensemaking of the practical and affective consequences for practitioners inhabiting and enacting that complexity.
3. Besides numerically based descriptions, abstracted explanations and procedural prescriptions, research evidence must encompass experiential data, collaboratively produced accounts and/or experience-based designs.

THE NORDIC CONTEXT

The Nordic countries consist of Denmark, Finland, Iceland, Norway and Sweden. The Nordic model is one of the welfare states characterised by a large public sector, welfare services and a social safety net. High standards of living, high levels of employment, relatively small differences in income, a high level of gender equality and the population's trust in public systems have been indicated as success factors (Nordic Council of Ministers, 2014). The Nordic model is also used in characterising the regulation of occupational safety and work environment in Nordic countries based on a three-pillar system involving employers, employees and regulatory bodies (Karlsen and Lindøe, 2006). By means of specific rules and procedures which demand the employer and employees to co-operate, mandatory performance standards and participative relations between workers and management, the Nordic countries have, since the 1970s, gradually improved the working environment giving enhanced safety for the workers.

The Nordic model is visible in the healthcare systems of the Nordic countries by publicly funded health services, social rights and equal access to health services. Globalisation, increasing life expectancy and economic trends, put pressure on the Nordic welfare states. A smaller number of people will have to support the ageing population, increasing the demand on public finances. Within the healthcare sector, more diseases will be treatable, and increasingly costly medical technology will be developed (Nordic Council of Ministers, 2014). Alongside these pressures, the social and health services need to remain stable, safe and of high quality. Co-operation between the Nordic governments on health falls under the auspices of the Nordic Council of Ministers on Health and Social Affairs. The Nordic co-operation has developed a strategy for the social affairs and health sector from 2013 and onwards, focusing on four specific objectives designed to strengthen sustainable welfare and health in the Nordic region:

1. Ensuring social security in the Nordic region in a constantly changing labour market
2. Active preventive measures
3. Improving quality and safety in the social and health services
4. Promoting innovation and research

In light of this book, the third objective on quality and safety is of specific relevance. Here, the Nordic countries will focus on exchanges of experience and knowledge in order to 'ensure that the services are effective, safe, co-ordinated, involve the users and exploit resources effectively' (Nordic Council of Ministers, 2013, p. 71). It is worth noting that the quotation covers five of the six quality dimensions suggested by the IOM in their common definition of quality in healthcare (IOM, 2000). For more information on the Nordic countries, see Chapter 1.

The patient rights focus in the Nordic countries is distinct and unambiguous with health legislation reinforcing policies ensuring equal access to health services. At the turn of the century, the Nordic region had undergone a rapid development in legally safeguarding patient rights compared to many other countries. Of eight European countries introducing such laws at the time, four were Nordic (Denmark, Iceland, Finland, Sweden) (Fallberg, 2000). Norway had their Patient Rights Act enforced from 2001. The Nordic policymakers have since then attempted to move the healthcare systems towards greater patient centredness by promoting patient rights; involving patient organisations; and encouraging patient choice, patient information and patient feedback. For more information on patient-centred care in the Nordic countries, see Chapter 2.

It is worth noting that the Nordic countries have not experienced any high-profile system failures such as the Bristol case or the Mid Staffordshire case that have driven the governmental regulation or research focus in a certain direction. With a population of approximately 26 millions (OECD, 2013), one might also claim that the Nordic countries are 'midgets' when it comes to research budgets and power to influence research agendas. Chapter 1 states that there are 163 research publications from the Nordic countries in peer-reviewed journals between 2000 and 2014. It should be of common interest to see how these research contributions give another angle at patient safety and quality in healthcare compared to other countries.

A NORDIC RESEARCH PERSPECTIVE?

An opportune question is whether or not it is possible to distil the characteristics of a Nordic perspective to research on patient safety and quality in healthcare. On the one hand, the Nordic countries are heavily influenced by international strategies and approaches, seen in, for example, the adoption of the IOM quality dimensions, WHO guidelines (e.g. safe surgical checklists) and national campaigns and accreditation programmes adopted based on the experiences of other countries. Based on the contributions of this book, we would claim that this adoption of international influence is more salient at a national macro-political and strategic level, while at an organisational meso-level and work practice–based micro-level, the Nordic traits are more apparent.

At the macro-level, we can trace international topics such as centralisation (volume–outcome relationship), accreditation and national campaigns in the Nordic research literature (see also Chapters 6 and 8). In Chapter 7, the authors document that the Nordic countries and in particular Sweden and Denmark have a long tradition for working with national clinical registries and therefore have extensive experience with both the strengths and limitations of such registries. This should be of particular interest for the increasing number of other countries planning to establish national clinical registers. We assume that country size is an explanatory factor for why some of the Nordic countries have succeeded in their work with national registries. On the contrary, the authors of Chapter 8 (Engel and Boje Andersen) speculate why the national strategy on hospital accreditation in Denmark has not been as successful as in other Nordic and non-Nordic countries. The emphasis on 'bottom-up' approaches refraining from 'forcing' individuals or departments into improvement efforts that is characteristic of the Nordic model might be an explanatory factor here.

At the meso-level, issues pertaining to a system perspective seem to characterise parts of the Nordic literature on patient safety and quality. Chapter 1 documents a number of studies trying to identify factors influencing patient safety and quality outcomes as well as understanding work processes and contexts. This might not be a very particular feature of the Nordic research perspective solely. Given the complex features of modern healthcare and its context sensitivity, such studies should nevertheless be valued in light of the specificities of the Nordic context which taken together with other contexts needs to be addressed and described if we are to effectively inform patient safety and quality improvement efforts.

At the micro-level, the perspective of patients and next of kin is highly evident in the Nordic healthcare policies (see Chapter 2) and in the patient safety and quality research literature (see Chapter 1). Policy wise, the patient perspective involves patient rights, transparency and access to information on system performance. Research wise, the patient perspective includes numerous accounts of patient experiences, their preferences and evaluations of the healthcare services they have received. We claim this is a cornerstone of the Nordic model and approach to healthcare provision. This entails principles of equality, the legal rights of patients and minimal power distance between providers and recipients of care characteristic of the Nordic welfarism and thinking.

A generic issue salient in both the international and Nordic research literature on patient safety and quality is the lack of explicit theory application and development (Chapters 1 and 4). The authors of Chapter 4 (Aase and Braithwaite) nevertheless claim that selected Nordic scholars and settings have had a profound influence on current research. These scholars are Jens Rasmussen (socio-technical approach to safety), Erik Hollnagel (self-organisation, organisational resilience), John Øvretveit (evaluation of quality improvement) and a setting Jönköping in Sweden (system-wide improvement). We can trace the contributions of these scholars and setting and see their ideas and influence having a discernible Nordic flavour, and as being reflected in current Nordic and international safety and quality research.

A READER'S MANUAL

Each of the book chapters can be read separately and without knowledge of the contents of the preceding chapters. The contributions in the book have been grouped into five different sections (introduction, perspectives, macro-political issues, meso-level organisational issues and micro-level work system issues) each of them forming a natural collocation useful for coherent reading.

In Chapter 1 (Husebø, Wiig, Guise, Storm, Sætre Hansen), an integrative systematic review has been conducted of the Nordic research literature on patient safety and quality (2000–2014) in order to conclude with suggestions for a future research agenda. Research needs to involve amongst others evaluations of strategies and initiatives for patient involvement, studies across primary care and specialist care (as opposed to single-level studies), cross-country comparisons and the study of the association between organisational and cultural characteristics and healthcare outcome.

In Section I, we present a set of generic issues relating to research on patient safety and quality. Storm and Coulter (Chapter 2) describe patient-centred care in the Nordic countries concluding that the topic is high on the macro-political agenda but there is still more to do to ensure that the healthcare systems are truly patient centred. Wiig and Manser (Chapter 3) reflect on the methodological challenges of researching patient safety and quality in healthcare, based on three characteristics of the healthcare context: conceptualisation of patient safety and quality, continuous contextual changes and patient involvement. Aase and Braithwaite (Chapter 4) indicate that theory seems to be under-represented in the international as well as the Nordic literature on patient safety and quality in healthcare but still claim that selected Nordic scholars have influenced the theoretical thinking within patient safety and quality improvement, respectively. Bovenkamp, Stoopendal and Bal (Chapter 5) delve into the topic of institutionally layered healthcare systems, using the case of the Netherlands as an example. Such institutional complexity can be seen also in the Nordic countries, demonstrating the need for addressing the institutional context of healthcare organisations and their power relations vis-à-vis external parties, hospital boards and professionals' organisations.

In Section II, we continue with raising macro-political issues of importance for patient safety and quality, presenting three Danish research contributions. Kjær Kristensen and Paaske Johnsen (Chapter 6) present an important overview of research studying the impacts of centralisation efforts on quality of care and costs, addressing the heavily debated volume–outcome relationship. Bartels, Mainz and Paaske Johnsen (Chapter 7) describe experiences with the use of national clinical registries as an essential component of a well-functioning healthcare system. Engel and Boje Andersen (Chapter 8) address the Danish strategy of using hospital accreditation as a means to improve patient safety and quality, reflecting on the spread of control requirements and a gap between perceived meaningful clinical quality and the control requirements.

In Section III, we concentrate on regional and organisational meso-level issues, presenting three contributions covering regional networks, co-ordination issues and management issues. Tervo-Heikkinen, Äijö and Holopainen (Chapter 9) describe the importance of establishing a multidisciplinary and multi-actor network for falls prevention in a Finnish region. Nedreskår and Storm (Chapter 10) study the hands-on effects of a national reform on cross-level co-ordination of discharge practices for elderly, documenting changes in management continuity, informational continuity and relational continuity. Bergerød and Wiig (Chapter 11) address how leadership along with organisational and cultural factors strongly influences patient safety and quality improvement work in two Norwegian hospital settings, stating how managers and their organisations have different responses dependent on their contextual setting.

In Section IV, we address a selection of micro-level issues pertaining to work systems and clinical practice. Guise, Husebø, Storm, Moltu and Wiig (Chapter 12) explore healthcare professionals' perspectives on patient safety and quality implications of implementing virtual home healthcare visits, indicating that positive implications are related to current Norwegian healthcare policy. Ekstedt (Chapter 13) describes how cognitive strategies are used by professionals in specialised home care settings to manage problem-solving and decision-making in everyday clinical work, using sensemaking theory to contribute to this understanding. Kivekäs, Haatainen, Koiki and Saranto (Chapter 14) address the issue of local variability using the case of intravenous medication errors with infusion pumps in different hospital settings.

Karina Aase
Lene Schibevaag

REFERENCES

Berg, R. and Cassells, J. (1992). *The Second Fifty Years: Promoting Health and Preventing Disability.* Washington, DC: National Academy Press.

Dutch Safety Board (2008). Een onvolledig bestuurlijk proces: Hartchirurgie in UMC St Radbound. Rapport Onderzoeksraad voor Veiligheid. Den Haag, April 2008.

Francis, R. (2013). *Report of the Mid Staffordshire NHS Foundation Trust Public Inquiry: Executive Summary.* London, U.K.: The Stationery Office: Mid Staffordshire NHS Foundation Trust Public Inquiry.

Hallberg, L.H. (2000). Patients rights in the Nordic countries. *European Journal of Health Law*, 7: 123–143.

Iedema, R. (2009). New approaches to researching patient safety. *Social Science & Medicine*, 69: 1701–1704.

IOM. (2000). *To Err Is Human: Building a Safer Health System.* Washington, DC: National Academy Press.

Jha, A.K., Prasopa-Plaizier, N., Larizgoitia, I., Bates, D.W., on behalf of the Research Priority Setting Working Group of the WHO World Alliance for Patient Safety. (2010). Patient safety research: An overview of global evidence. *Quality and Safety in Health Care*, 19: 42–47.

Karlsen, J.E. and Lindøe, P. (2006). The Nordic OSH model at a turning point? *Policy and Practice in Health and Safety*, 14(1): 17–30.

Kennedy, I. (2001). The report of the public inquiry into children's heart surgery at the Bristol Royal Infirmary 1984–1995: Learning from Bristol. Presented to Parliament by the Secretary of State for Health, July 2001.

Nordic Council of Ministers. (2013). Nordic co-operation on social affairs and health. Strategy for the social affairs and health sector 2013 and onwards. ANP 2013:750. http://dx.doi.org/10.6027/ANP2013-750 (accessed 14 November 2015).

Nordic Council of Ministers. (2014). Does the Nordic model need to change? Overview of the research report, 'The Nordic model – Challenged but capable of reform'. ANP 2014:772. http://dx.doi.org/10.6027/ANP2014-772 (accessed 14 November 2015).

OECD. (2013). Health at a glance, OECD indicators. OECD Publishing, Paris, France. http://www.oecd.org/els/health-systems/Health-at-a-Glance-2013.pdf (accessed 14 November 2015).

Pronovost, P., Ravitz, A.D., Stoll, R.A. and Kennedy, S.B. (2015). Transforming patient safety. A sector-wide systems approach. Report of the WISH Patient Safety Forum 2015. *World Innovation Summit for Health*, Doha, Qatar.

Vincent, C. (2010). *Patient Safety*, 2nd edn. Oxford, U.K.: Wiley Blackwell.

Wiig, S., Aase, K., Plessen, C., Burnett, S. et al. (2014). Talking about quality: Exploring how 'quality' is conceptualized in European hospitals and healthcare systems. *BMC Health Services Research*, 14: 478.

World Health Organization. (2008). WHO guidelines for adverse event reporting and learning systems. From information to action. World alliance for patient safety, Worls Health Organization, Geneva, Switzerland.

1 Status of Nordic Research on Patient Safety and Quality of Care

Sissel Eikeland Husebø, Siri Wiig, Veslemøy Guise, Marianne Storm and Britt Sætre Hansen

CONTENTS

INTRODUCTION

This chapter provides an overview of the Nordic research on patient safety and quality based on a review of the literature. Evidence is summarised and synthesised to provide suggestions for a possible future Nordic research agenda.

BACKGROUND

In an international context, large-scale system failures such as the Bristol Royal Infirmary Inquiry (Kennedy, 2001, Kohn et al., 2000), the Mid Staffordshire Public Inquiry (Francis, 2013) and investigation of critical incidents in Dutch hospitals (Behr et al., 2015) have put patient safety on the agenda (Martin and Dixon-Woods, 2014). In all cases, large numbers of patients died or suffered due to insufficient care or malpractice over time and this practice was able to continue despite regulatory systems and inspectorates being in place to detect such failures (Weick and Sutcliffe, 2003). Compared to the Nordic countries, there have not been any system failures of a similar scale. Denmark, Finland and Sweden have specific patient safety laws, while Norway has incorporated patient safety and quality dimensions into several laws and regulations. Quality improvement strategies have been in place in the majority of the Nordic countries over several decades (since 1995 in Norway, 1990 in Sweden, 1975 in Denmark and 1994 in Finland). Iceland has a national quality development plan for the health services aiming at enhancing patient safety and quality and in 2007 the authorities published the first policy on quality in healthcare services (Sigurgeirsdóttir et al., 2014). The progress of patient safety and quality research in the Nordic countries has been more sporadic and fragmented, and to date, a limited overview exists.

To our knowledge, the Nordic research literature on patient safety and quality has not been reviewed earlier, and it is therefore worth undertaking further investigation to suggest a future Nordic research agenda in the Nordic countries.

AIM AND RESEARCH QUESTIONS

The aim of the current literature review is to provide an overview of the Nordic research literature on patient safety and quality in healthcare and to suggest directions for a future research agenda.

The review questions addressed are as follows:

1. What is the current research on patient safety and quality in healthcare?
2. Which empirical fields and research methods have been addressed?
3. Which domains of patient safety and quality can be identified?

THE NORDIC REGION: WHAT IS IT?

The Nordic region consists of the countries Denmark, Finland, Sweden, Iceland, Norway and the self-governed areas of the Faroe Islands, Greenland and Åland. The total population in the Nordic region is around 26 million inhabitants (Table 1.1).

TABLE 1.1
Population Based on Numbers from OECD (2013) and www.norden.org and Health Expenditure in the Nordic Countries

Country	Population in 2011 (OECD, 2013) (in thousands)	Health Expenditure per Capita 2011 (USD)	Health Expenditure as Share of GDP (%)
Norway	4.952	5669	9.3
Denmark	5.571	4448	10.9
Finland	5.388	3374	9.0
Iceland	319	3305	9.0
Sweden	9.447	3925	9.0

Source: OECD, *Health at a Glance 2013: OECD Indicators*, OECD Publishing, 2013. http://dx.doi.org/10.1787/health_glance-2013-en.

The region as a whole is sparsely populated, with the exception of Denmark (www.norden.org). The health expenditure figures for all five countries in the region are shown in Table 1.1 (OECD, 2013).

NORDIC CO-OPERATION TO PROMOTE PATIENT SAFETY AND QUALITY

The Nordic region has established a cross-country co-operation within health and social affairs. The co-operation falls under the responsibility of the Nordic Council of Ministers on Health and Social Affairs. The Nordic collaboration is anchored politically, financially and culturally and is an important player in a European and international context. The co-operation is based on the shared set of Nordic values, which constitutes the basis for the Nordic welfare model (Norden, 2013). The traditional image of the Nordic model is of a welfare state characterised by a large public sector that provides its citizens with generous benefits, welfare services and a social safety net (Norden, 2014). The welfare state is supported by the principles of equal opportunities and social solidarity and security for all inhabitants, regardless of gender, ethnicity, faith, belief, disability, age and sexual orientation. The welfare model promotes social rights and equal access to social and healthcare services, education and culture. This also applies to care of those who are part of socially disadvantaged and vulnerable groups in society (Norden, 2013, p. 65).

A common Nordic strategy on health and social affairs has been developed (Norden, 2013). The strategy includes the specific goal of improving patient safety and quality of health and social services. The strategy emphasises the common challenge of an ageing population with an increased need for care and advocates for experience and knowledge exchange to ensure quality.

During the past years, most hospital services in the Nordic countries have undergone centralisation processes based on a rationale of the relationship between organising, volume and quality of treatment. The strategy also focuses on Nordic co-operation to promote patient safety. Due to their size, the Nordic countries should work closer in clinical multi-centre studies to evaluate the effectiveness and safety of new diagnostic methods and treatment. The development and use of new technology and e-health is also a prioritised area (Norden, 2013). User involvement and patient perspectives are furthermore a critical part of the Nordic strategy to achieve high-quality care, and patient experiences are regularly followed up by collecting standardised national patient surveys in specialist healthcare settings (Anell et al., 2012; Holmboe et al., 2014). The extent to which these data are used to guide quality improvement and research in the clinical setting is however not clear.

METHODS

Design

An integrative systematic review methodology was used (Burns et al., 2011, pp. 418–463). This involves a multi-stage strategy that includes problem identification, literature search, data evaluation, data analysis and presentation (Whittemore and Knafl, 2005). The chosen approach was considered appropriate to allow for inclusion of studies with diverse methodologies and provide a fuller understanding of current research in the Nordic countries (Whittemore and Knafl, 2005). To minimise bias in the review process, a review protocol was developed, and a systematic search in line with Lefebvre et al. (2011) was undertaken.

Search Methods

Two of the authors (VG and SEH) searched the following five online databases in the period 1 June to 31 July 2014: Cinahl, ASP, Medline, Scopus and PubMed. The databases were chosen for their comprehensive collection of peer-reviewed academic journals in the area of patient safety and quality of care. The search was conducted using a combination of the following terms: *patient safety*, *quality*, *quality improvement*, *Nordic*, *Norway*, *Sweden*, *Denmark*, *Finland* and *healthcare*. Table 1.2 displays the databases, search terms, combinations and number of hits.

The search was limited to information available in the title, abstract and keywords. Items had to feature empirical data material from at least one of the Nordic countries to be included. Quality assessment was not undertaken, and therefore, no items were excluded based on the assessment of the methods, study design or outcomes. Further inclusion and exclusion criteria were as follows:

- *Language*: abstracts in English, Norwegian, Danish, Swedish or Finnish were included.
- *Date*: only studies published between 2000 and 2014 were included.
- *Clinical setting*: studies from both primary and specialised healthcare services were included.
- Abstracts from study protocols, PhD theses and research conferences were excluded.

TABLE 1.2
Search in Databases, Terms and Hits

Databases	Terms and Combinations	Hits
Cinahl	Patient safety	10,409
ASP	Patient safety	52,249
Medline	Patient safety	47,529
Scopus	Patient safety	62,391
Pubmed	Patient safety	14,166
Cinahl	Quality/quality improvement	81,243
ASP	Quality/quality improvement	403,075
Medline	Quality/quality improvement	412,666
Scopus	Quality/quality improvement	1,714,996
Pubmed	Quality/quality improvement	57,114
Cinahl	Healthcare	29,433
ASP	Healthcare	48,446
Medline	Healthcare	74,746
Scopus	Healthcare	172,404
Pubmed	Healthcare	85,062
Cinahl	Nordic/Norway/Sweden/Denmark/Finland	8,945
ASP	Nordic/Norway/Sweden/Denmark/Finland	57,168
Medline	Nordic/Norway/Sweden/Denmark/Finland	4,705
Scopus	Nordic/Norway/Sweden/Denmark/Finland	186,530
Pubmed	Nordic/Norway/Sweden/Denmark/Finland	52,886
Cinahl	Combined search	92
ASP	Combined search	176
Medline	Combined search	288
Scopus	Combined search	984
Pubmed	Combined search	303
Total number of hits		1,843
Duplicates removed		801
Excluded		8
The total number of hits screened		1,034

Review Process

A two-phased literature review process was conducted. In phase 1, we applied a broad approach to answer review questions one and two. In phase 2, a framework for patient safety research and improvement (Pronovost et al., 2009) was used to obtain a deeper analysis of the included abstracts to answer review question three.

In phase 1, 1034 abstracts were divided among the five chapter authors and a team of researchers from the 'Quality and Safety in Healthcare Systems' research group at the University of Stavanger (see 'Acknowledgement' section). Each reviewer then worked individually to include or exclude items according to information in the title, abstract and keywords and the inclusion/exclusion criteria. During this process, the reviewers excluded 844 of 1034 abstracts from the sample because they did not meet the inclusion criteria. All reviewers used a data extraction tool to get a clear overview of the remaining 190 studies, while ensuring further consistency in the review process. The data extraction tool covered the dimensions of (1) author, year of publication and country; (2) title; (3) theory used; (4) methods used; (5) empirical field; (6) topics (limited to five keywords) and (7) any specific issues worth noting.

In phase 2, authors BSH and SEH closely assessed the remaining 190 abstracts. Based on the inclusion criteria and the research questions, 27 additional abstracts were removed during this process, which resulted in a sample of 163 abstracts included for data extraction and further analysis according to a framework for patient safety research and improvement (Pronovost et al., 2009).

Analytical Framework

Pronovost et al.'s (2009) 'Framework for Patient Safety Research' was used to classify the included abstracts. The framework includes five domains with associated subdomains (see Table 1.3). The framework was developed to address many of the issues emerging from a growing international desire for higher quality and safer

TABLE 1.3
Framework for Patient Safety Research

Domain and Subdomains	Description
1. Evaluating progress in patient safety	Develop valid and feasible measures to evaluate progress to improve patient safety
2. Translating evidence into practice A. Summarise the science B. Measure performance C. Understand the current process and context of work D. Ensure all patients reliably receive the intervention and patient/next of kin experiences	Develop and evaluate interventions that increase the extent to which patients receive evidence-based medicine
3. Assessing and improving culture A. What is safety culture? B. How do you measure safety culture? C. How do you use safety culture results?	Strategies and interventions to improve safety culture and communication
4. Identifying and mitigating hazards A. Retrospective identification of hazards B. Prospective identification of hazards C. Mitigating hazards	Use of retrospective and prospective analyses to identify and mitigate safety hazards at the microscopic level (unit or department, in-depth evaluation) and macroscopic (institutional, country) levels
5. Evaluating the association between organisational characteristics and outcomes	Evaluate organisational characteristics that help or hinder research efforts or patient safety practices, for example the association between staffing and patient outcomes.

Source: Pronovost, P. et al., *Circulation*, 119, 330, 2009.

healthcare and is based on the Institute of Medicine's strategies for improvement and literature regarding knowledge transfer and diffusion of innovation (IOM, 2001; Pronovost et al., 2009; Stepnick and Findlay, 2003). Quality studies are included in the framework, since the authors acknowledged that the boundaries between safety and the broader concept of quality remain poorly defined (Pronovost et al., 2009). We considered various analytical frameworks as Donabedian (2005) as others have done (e.g. Jha et al., 2010). However, Pronovost et al.'s (2009) framework was chosen because it provides more detailed information about the illuminated topics beyond.

Each abstract was read several times and coded, and then coding judgements were discussed openly. Coding accuracy within each domain and subdomains was done in line with Pronovost et al. (2009). In domains where subdomains were insufficiently described in Pronovost et al. (2009), abstracts were coded based on the two authors' (BSH and SEH) own interpretation.

RESULTS

A main finding in this Nordic literature review is the large amount of abstracts involving qualitative studies focusing on patient perspective, patient involvement, and patients' and next of kin's experiences with healthcare services. Most of the Nordic literature on patient safety and quality in healthcare published in peer-reviewed journals comes from Sweden. The methodological approaches vary and include a mix of quantitative and qualitative methods. There is an overweight of literature from the specialised healthcare services, with much less attention on primary care services. The included abstracts cover all domains in the Framework for Patient Safety Research (Pronovost et al., 2009), with most abstracts (58) belonging in domain 2 – Translating evidence into practice, while fewest abstracts (8) were allotted to domain 5 – Evaluating the association between organisational characteristics and outcomes. In the remainder of the results section, we present the detailed findings of the two-phased review process.

Phase 1: Distribution between Countries, Theoretical, Empirical and Methodological Contributions

The distribution of abstracts between countries is as follows: Norway, 35; Denmark, 33; Sweden, 77 and Finland, 18. Only 8 of the 163 included items mention the use of theory in the abstract, including implementation theory and models for understanding success in quality improvement, organisational learning, indicators, guidelines and standardisation. Empirical fields are summarised in Table 1.4 according to specialised healthcare, primary healthcare, a mix of primary and specialised healthcare and others.

The amount of abstracts in the 'other' category is high. Several abstracts were difficult to review with regard to empirical field. The 'other' category includes studies focusing on data from national registries, evaluation of guidelines, reviews, indicators, waiting lists, presentation of national projects, reimbursement and several studies stating 'healthcare' as empirical setting.

Abstracts from the specialised healthcare services involve in-hospital, pre-hospital, maternity, intensive and emergency care and education and training of

TABLE 1.4
Empirical Fields Covered in the Nordic Patient Safety and Quality Literature

Empirical Field	Distribution of Abstracts
Primary healthcare	30
Specialised healthcare	72
Primary and specialised healthcare	7
Other	54

TABLE 1.5
Methodological Approaches Used in Nordic Patient Safety and Quality Literature

Methodological Approach	Distribution of Abstracts
Qualitative	53
Quantitative	88
Mixed	7
Review	8
Not stated	7

health professionals. Abstracts from primary care setting cover nursing homes, home healthcare services, general practitioner services and dental care in nursing homes.

The assessment of methodological approaches indicates a big variety of qualitative, quantitative and mixed methods being used (see Table 1.5). The qualitative methods include case studies, semi-structured interviews, focus group interviews, observation studies, action research, studies of root cause analyses, retrospective patient record reviews and document analysis.

The quantitative methods include surveys and questionnaires, randomised controlled trials, national registries studies, evaluation of guidelines, pre-test and post-test measurement, cohort studies and prospective cohort studies.

Among the included abstracts are eight literature reviews. Topics here cover impact of work culture on quality in nursing homes, oral hygiene in elderly people in hospitals and nursing homes, an automated dose dispensing service for primary healthcare patients, a road map for patient safety research approaches, costs of patient safety and quality in healthcare, research on quality improvement and quality evaluation and indicators and quality data based on national clinical databases.

Phase 2: Current Domains of Patient Safety and Quality Research in the Nordic Setting

We found that the largest proportion of abstracts (58) cover domain 2 – Translating evidence into practice, including 38 abstracts on patient/next of kin experiences, followed by domain 1 – Evaluating progress in patient safety (48), domain 4 – Identifying and mitigating hazards (31), domain 3 – Assessing and improving culture (18) and domain 5 – Evaluating the association between organisational characteristics and outcomes (8) (Pronovost et al., 2009) (see Table 1.6). A list of the 163 journal articles coded and analysed in the review appears as Appendix A.

Domain 1 – Evaluating Progress in Patient Safety (48 Studies)

Most of the abstracts evaluating progress in patient safety focus on the development and testing of patient safety and quality measurements and indicators and factors that influence patient safety and patient outcomes. A few abstracts are concerned with the monitoring of healthcare-associated infections, and a few deal with the development of theory, models and methods.

Development and testing of measurements and indicators (21 studies). Several studies report development and testing of a variety of measurements, such as health information and communication technology (ICT) (Adler-Milstein et al., 2014), process and outcome improvements in healthcare (Andersson et al., 2013), a reporting system for patient safety incidents (Bjørn et al., 2009), risk assessment concerning pressure ulcers (Fossum et al., 2012), patient participation in the Emergency Department (Frank et al., 2011), Global Trigger Tool (Mattsson et al., 2013; Schildmeijer et al., 2012), the Nordic Patient Experiences Questionnaires (Oltedal et al., 2007; Skudal et al., 2012), a cross-cultural measure to assess parents' satisfaction with care of children with chronic conditions (Schmidt et al., 2008) and a retrospective record review of no-harm incidents to improve patient safety (Schildmeijer et al., 2013). Development and testing of indicators and checklists comprise patient safety indicators, checklist for outpatient surgery (Helmiö et al., 2012), best practice in the delivery of healthcare to immigrants (Jensen et al., 2010), clinical indicators and standards in an outpatients' department for sub-acute low back pain patients (Johansen et al., 2004), patient-centred healthcare quality indicators for rheumatoid arthritis (Petersson et al., 2014), evidence-based national indicators (Wettermark et al., 2006b), patient-specific quality indicators and quality evaluation and indicators for tonsillectomy (Sjøhart Lund et al., 2010).

Factors that influence patient safety and patient outcomes (20 studies). Studies on influencing factors include, for example, physician's self-evaluation of the consultation (Ahlén and Gunnarsson, 2013), home healthcare nurses awareness of fall prevention in clinical practice (Berland et al., 2012) and that the selection of older patients eligible for a surgical treatment is likely to be based on the health status and the safety of surgical procedures rather than chronological age (Oksuzyan et al., 2013).

Studies on quality improvement methods are common and positive changes are reported concerning clinical guidelines and communication methods such as Situation – Background – Assessment – Recommendation (Wallin and Thor, 2008).

A few studies emphasise using ICT as a tool to improve patient safety in healthcare. The studies demonstrate a demand for a common exchange format, a need for ICT tools for documenting clinical contents, significant gaps in design, implementation and use of ICT in healthcare to ensure patient safety and that a change from paper forms to electronic laboratory requests by general practitioners show

TABLE 1.6
Domains and Subdomains Covered in the Nordic Patient Safety and Quality Literature

Domains and Subdomains	No. of Studies	Study ID
1. Evaluating progress in patient safety	48	3,6,11,13,14,16,17,20,27,32,35,37,44,48,49,50, 53,62,64,65,72,75,80,81,86,88,91,94,98,111,112, 113,119,121,124,127,131,134,135,140,145,150, 151,153,159,160,161,162
A. Summarise the science	1	13
B. Development and testing of measurements and indicators	21	3,11,20,32,35,37,48,62,64,75,88,94,98,111,112,114, 119, 121,140,150,161
C. Factors that influence patient safety and patient outcome	20	6,14,16,17,27,49,65,72,80,81,86,91,113,131,135,145, 151,153,159,160
D. Monitoring healthcare-associated infections	3	44,127,134
E. Development of theory, models and methods	3	50,53,162
2. Translating evidence into practice	58	2,5,10,19,22,23,26,30,31,36,37,39,40,41,42,43, 55,57,59,60,63, 66,69,71,73,74,77,84,87,89,93, 99,102,103,104,105,107,108,110,117,118,120, 122,128,130,132,133,137,141,142,143,146,147, 149,154,156,157,163
A. Summarise the science	1	117
B. Measure performance	2	99,110
C. Understand the current process and context of work	17	22,23,30,31,57,60,63,71,73,102,103,107,143,146,147, 149,163
D. Ensure all patients reliably receive the intervention and patient/next of kin experiences	38	2,5,10,19,26,36,38,39,40,41,42,43,55,59,66,69,74,77, 84,87,89,93,104,105,108,118,120,122,128,130,132, 133,137,141,142,154,156,157
3. Assessing and improving culture	18	1,28,34,47,54,61,68,82,92,90,96,97, 115,123,124,129,136,144
A. Theory and measurement development	2	47,54
B. Mapping patient safety culture	4	1,61,92,97
C. Factors that influence communication	12	28,34,68,82,90,96,115,123,124,129,136,144
4. Identifying and mitigating hazards	31	8,9,15,18,21,24,25,45,46,52,56,58,67,70,76,78,79, 83,85,95,100,101,106,109,116,125,126,138,139, 152,158
A. Medical hazards	21	8,9,18,21,24,25,52,70,78,83,85,95,100,101,106,109, 116,125,138,139,158
B. Medication hazards	10	15,45,46,56,58,67,76,79,126,152
5. Evaluating the association between organisational characteristics and outcomes	8	4,7,12,29,33,51,148,155

positive effects on reducing errors and costs (Bernstein et al., 2006; Johansen and Rasmussen, 2009; Kuusela et al., 2011).

Only one study used risk analysis to develop and manage routines for reducing the risks of an ICT tool and found a reduction in risk due to technology and equipment and in the area of training and competence (Öhrn and Eriksson, 2007). Studies in medication procedures report a need for improved communication and information in primary care (Bakken et al., 2007), whereas variation between the different primary care centres with regard to the prescribing doctors' compliance with guidelines from the regional drug and therapeutics committee was found to be great (Wettermark et al., 2006a). The utilisation of drugs was difficult to forecast due to uncertainties about the rate of adoption of new medicines and various ongoing healthcare reforms and activities to improve the quality and efficiency of prescribing (Wettermark et al., 2010).

Monitoring healthcare-associated infections (3 studies). Only a handful of studies examine, for example, healthcare-associated infections (Hajdu et al., 2011; Struwe et al., 2006; Thorstad et al., 2011).

Development of theory, models and methods (3 studies). This subdomain comprises only three studies focusing on the development of theory and methods (Hofoss and Deilkås, 2008; Hovlid and Bukve, 2014; Øvretveit, 2002).

Domain 2 – Translating Evidence into Practice (58 Studies)

The studies included in the domain of translating evidence into practice are varied in setting and perspective. Most studies highlight experiences and perceptions of healthcare services from patients and next of kin. Many other studies focus on healthcare workers' perspectives on safety and quality issues.

Ensure all patients reliably receive the intervention (38 studies). This subdomain has the largest amount of studies included. The studies focus, for example, on the effects of an educational programme for heart failure patients (Agvall et al., 2013), the need for improvement in adherence to guidelines for medication in long term and asthma treatments for children (Ingemansson et al., 2012), how implementation of ICT can influence documentation practice regarding pressure ulcers and malnutrition in nursing homes and hospitals (Fossum et al., 2013; Gunningberg et al., 2009) and examination of nurses' knowledge and knowledge gaps on the prevention and treatment of pressure ulcers (Källman and Suserud, 2009). Several intervention programmes are examined, such as one on patient rehabilitation (Frølich et al., 2010), another on follow-up of older patients after discharge (Rytter et al., 2010), a multi-professional teamwork programme in primary care (Sipilä et al., 2008) and a quality improvement programme which demonstrates better patient outcomes (Thor et al., 2010).

The studies dealing with evaluation, experience and perception of healthcare services from the patients and next of kin perspective contribute to extensive knowledge in this subdomain. The studies reveal ambiguous findings concerning experiences and preferences of care and treatment. On the one hand, some studies feature hospitalised patients expressing satisfaction with involvement and participation in decision-making and the quality of care and treatment (Browall et al.,

2013; Skarstein et al., 2002). On the other hand, shortcomings in meeting patients' needs and expectations, in undertaking patient-centred care, and in patient involvement in patient safety are reported in other studies (Bjorkman and Malterud, 2009; Kaakinen et al., 2013; Wolf et al., 2012). One study also noted ambiguous results regarding whether or not communication training programmes for healthcare professionals influence patient outcomes (Nørgaard et al., 2012).

Several studies reporting quality evaluations from parents and relatives reveal shortcomings in paediatric care and patient transfer within hospitals and mental health care, but there are also reports of satisfaction with the quality of maternity care as evaluated by immigrant parents (Ammentorp et al., 2001; Häggström et al., 2014; Johansson et al., 2014; Ranji et al., 2012).

Understand the current process and context of work (*17 studies*). This subdomain incorporates results regarding the perceptions, experiences and responses of nurses and physicians regarding 'good' patient care, patient education and medical treatment, prerequisites for implementing evidence-based practice and factors promoting quality improvement. Prerequisites for the implementation of guidelines in fast-track surgery and standardised care plans are involvement of nurses in the process and use of a workshop-practice method (Jakobsen et al., 2014; Jakobsson and Wann-Hansson, 2013). Implementation and effects of ICT reveal differences in nurses' and doctors' reporting versus improvement of outpatient referral and cost efficacy (Rahimi et al., 2008, 2009).

Summarise the science (*1 study*) which is a literature review showing that patients using an automated dose dispensing service had more inappropriate drugs in their regimens (Sinnemäki et al., 2013).

Measure performance (*2 studies*). In this subdomain, we included only two studies, one regarding which method has the best effect on cardiopulmonary resuscitation outcomes (Putzer et al., 2013) and one if a specialised cancer centre can achieve good local control on patients with inadequate surgery (Sampo et al., 2012).

Domain 3 – Assessing and Improving Culture (18 Studies)

In the domain of assessing and improving culture, we include studies assessing safety culture or examinations of factors that influence communication.

Theory and measurement development for assessing safety culture (*2 studies*). This subdomain comprises only two studies focusing on theory development (Hovlid et al., 2012) and psychometric properties of the 'Hospital Survey on Patient Safety Culture' measurement tool (Hedsköld et al., 2013).

Mapping patient safety culture (*4 studies*). Two studies mapping patient safety culture (Aase et al., 2008; Olsen and Aase, 2010) were included. The other two included studies give insights into patients' views on adverse events and older patients' perspectives on medical approaches (Itoh et al., 2006; Pedersen et al., 2008).

Factors that influence communication (*12 studies*). In this subdomain, the included studies deal with a variety of factors that influence communication between healthcare providers and patients. Listening to the patient views through a structured computer-facilitated patient–clinician dialogue focuses treatment sessions and has a positive effect on patients' quality of life, need for care, and treatment satisfaction

(Bylund, 2008), using patients' knowledge may improve education materials provided to patients undergoing colorectal cancer surgery (Smith et al., 2014) and using cognitive artefact analysis improve patient handovers between hospital and community care (Johnson et al., 2013). In primary care settings, factors that influence communication include hospital information to patients with a need for follow-up by general practitioners (Vægter et al., 2012).

Domain 4 – Identifying and Mitigating Hazards (31 Studies)

In this domain, most of the included studies identify adverse events, medication errors and mortality rates in the hospital setting.

Medical hazards (21 studies). Several of the studies identify adverse events (complications and infections) and mortality in the hospital setting. These include post-operative complications after, for example, incontinence and urinary stress incontinence surgery (Ammendrup et al., 2009a,b), in conjunction with temporary pacemaker treatment (Bjørnstad et al., 2012), vascular surgery (Brattheim et al., 2011), orthopaedic care (Unbeck et al., 2010), cancer treatment and care (Lipczak et al., 2011) and intracranial tumour operations (Solheim et al., 2012). Communication failure between staff members is another important contributing factor of severe patient safety incidents (Rabøl et al., 2011). For orthopaedic patients, many more adverse events are detected using the Wimmera model with manual screening than in other traditional local and nationwide reporting systems (Unbeck et al., 2008). Patients dying of heart disease compared to those dying from cancer were found to have received lower quality of care (Brännström et al., 2012). In emergency departments, adverse events were mainly found to be related to overcrowdings, lack of beds on wards and intensive care units, technical problems, diagnostic procedures and treatments (Khorram-Manesh et al., 2009). In Danish mental healthcare, psychiatric patients were found to receive a lower quality of somatic care (Steinø et al., 2013).

In telenursing, threats to patient safety were related to the surrounding society, organisation of telenursing, the telenursing practice and the caller (Röing et al., 2013). In primary care, the presence of subjective memory complaints for older people was not significantly associated with an increase in all-cause mortality (Siersma et al., 2013).

Medication hazards (10 studies). Several included studies report medication errors and factors that influence such errors. Studies from the hospital setting reveal that interruption during the preparation of medication increases the risk for errors (Berg et al., 2013; Härkänen et al., 2013), the same applies to the high number of generic drugs and frequent generic substitutions in use (Håkonsen et al., 2010). In one study, 20% of drug users reported adverse drug events during the previous month (Hakkarainen et al., 2013). Interventions such as repeated adverse drug reactions information letters to physicians and nurses did not increase the reporting rate of adverse drug events (Johansson et al., 2011). Similarly, the use of electronic patient record systems and tools did not change work processes related to medication procedure and information management to a safer procedure (Kivekäs et al., 2014). In primary care, gross variations in general practitioners' practice exist regarding medication treatment (Kristoffersen et al., 2006).

Domain 5 – Evaluating the Association between Organisational Characteristics and Outcomes (8 Studies)

Only a handful of studies examine the association between organisational characteristics and outcomes, and the themes and findings of this domain diverge in several directions. For example, included studies in specialised healthcare settings report that specially educated healthcare staff may contribute to good quality diabetes care (Adolfsson et al., 2010), identify important factors related to the assessment of patient safety (Alenius et al., 2014), report associations between work culture and quality of care (André et al., 2013), note a discrepancy between the actual complaint rate of patients' complaints of adverse events and the number of respondents stating that they have had reasons to complain (Wessel et al., 2012), discuss how waiting lists or queues could be reduced (Eriksson et al., 2011) and demonstrate that healthcare policymakers and hospital leaders have to design and implement strategies to help their staff members recognise and value the contribution that patient involvement and patient experiences can make to the improvement of healthcare (Wiig et al., 2013).

In primary care, studies highlight delegation to unlicensed personnel as prerequisites for a functioning organisation involving a number of contradictions such as nurses lack of control, powerlessness, vagueness regarding responsibility and resignation (Bystedt et al., 2011) and note that district nurses are an underused resource regarding their work situation after the free-choice system in primary care (Hollman et al., 2014).

DISCUSSION

The aim of this review was to provide a status of the Nordic patient safety and quality research literature and to suggest directions for a future research agenda. We found that the main strength of the Nordic literature on patient safety and quality is the amount of studies of patients' and next of kin's perspectives (including experience, preferences and evaluations) on healthcare services. These findings do however reflect some ambiguous results regarding perceptions and experiences of meeting patients' needs and expectations, patient-centred care and patient involvement. This implies a need for rigorous multisite evaluations of the implementations of patient-centred care and patient involvement in the Nordic countries. One explanation for this feature of the Nordic research may be found in the Nordic healthcare model, characterised by short distances within healthcare systems hierarchies, and based on an ethos of strong patient rights and equality of healthcare services (Norden, 2013, 2014).

Among the Nordic countries, Sweden stands out with the largest proportion of studies, especially concerning large studies of patient registry data on, for example mortality rates in psychiatric healthcare and hospital-related adverse events (Björkenstam et al., 2012; Khorram-Manesh et al., 2009). The results from these two studies (Björkenstam et al., 2012; Khorram-Manesh et al., 2009) may give directions for future targeted interventions to decrease mortality rates in psychiatric healthcare and adverse events in hospitals in Sweden and across the Nordic countries.

Finnish researchers have done considerable work in identifying and mitigating medication risk in the hospital setting, indicating a need for the implementation of interventions to prevent and mitigate such errors. Several of the included studies have developed and tested measurements and indicators in a Nordic context. Such specific measures are needed to provide a sufficiently broad view on patient safety and quality and are a prerequisite to move the science forward (Pronovost et al., 2009). Consequently, research in this domain should continue.

Based on the reviewed abstracts, the three largest shortcomings in Nordic studies relate (1) performance measurement (process or outcome) in the healthcare setting, (2) assessing and improving safety culture and (3) evaluating the association between organisational characteristics and outcomes. To close these gaps, the Nordic countries have to continue to keep creating capacity in terms of organisational and cultural factors and methods to assess and measure performance in patient safety and quality research. Moreover, there is a need to create a research infrastructure between the Nordic countries and to evaluate the cost–benefit ratio of improvement efforts and interventions in line with suggestions from Pronovost et al. (2009) focusing on a much closer collaboration between researchers and hospital leaders.

There are no studies involving cross-country comparisons of patient safety and quality in the Nordic countries (Appendix A). This stands out as an important area for future research and corresponds well to the Nordic strategy on the health and social sector, emphasising the need for cross-country research within the Nordic countries in the years to come (Norden, 2013).

Another gap identified is the small amount of research abstract from the primary care setting or a combination of the specialised and primary care settings. The lack of patient safety research from a primary care perspective mirrors findings from the international literature (Pearson and Aromataris, 2009), adding emphasis to importance of knowledge in this area. Furthermore, only one identified study focuses on intervention sustainability, implying more attention to longitudinal studies in patient safety and quality both within and across the Nordic countries.

The review has several limitations that need to be addressed. The possibility of excluding relevant studies due to too broad or too narrow a search strategy (e.g. terms and databases used) is a potential threat to the study's validity. Data extraction was based only on the reading of abstracts, and neither quality appraisal nor hand searches in reference lists were not performed, which might also represent a threat to the validity of the review. In addition, the classifying of the included studies in phase 2 was done by only two of the authors, representing a risk of potential bias. A possible strength with the review is that the analyses identified several new subdomains in the patient safety framework of Pronovost et al. (2009). The review has also provided a broad overview of the patient safety and quality research in the Nordic countries.

CONCLUSION

The majority of the Nordic research on patient safety and quality relates to evaluating progress in patient safety, translating evidence into practice and identification of hazards. The 'patient' voice in the Nordic research literature contributes with valuable knowledge to the international patient safety and quality research community on how to further develop patient-centred healthcare services. Development and testing of measurements and indicators for patient safety is important in the continuing development of robust measures. The findings that identify and mitigate medical hazards can be used to learn how to prevent risks and adverse events.

There is a lack of studies addressing assessing and improving culture and evaluating the association between organisational characteristics and outcomes indicating a need for future research to assess the influence of organisational and cultural factors in patient safety and quality in the Nordic healthcare context. Our suggestions for the direction of future patient safety and quality research in the Nordic countries imply

- Rigorous multisite evaluations studies of implementations of patient-centred care initiatives and patient involvement strategies
- Large studies of patient registry data
- Identification and mitigation of medication-related risks studies
- Studies in primary care or a combination of specialised and primary care settings
- Studies with cross-country comparisons of patient safety and quality
- Studies of measuring performance
- Studies measuring performance, assessment and improvement of patient safety culture
- Evaluations of the association between organisational characteristics and healthcare outcomes

ACKNOWLEDGEMENTS

At *The Third Nordic Conference on Research in Patient Safety and Quality in Healthcare – NSQH 2014*, a pre-conference workshop was organised in Stavanger with the topic 'Nordic research on patient safety and quality in health care – status and suggestions for future direction'. We want to acknowledge and thank the scientific committee of the NSQH 2014 and the international group of experts for their contribution and valuable comments in the pre-conference workshop. Especially, we want to thank Tanja Manser, Jeffrey Braithwaite and Angela Coulter. We also want to thank members of the research group, Lene Schibevaag, Cecilie Haraldseid, Randi Nisja Heskestad and Dagrunn Nåden Dyrstad, for their contribution in review phase 1.

REFERENCES

Aase, K., Høyland, S., Olsen, E., Wiig, S. and Nilsen, S.T. 2008. Patient safety challenges in a case study hospital – Of relevance for transfusion processes? *Transfusion and Apheresis Science*, 39, 167–172.

Adler-Milstein, J., Ronchi, E., Cohen, G.R., Winn, L.A.P. and Jha, A.K. 2014. Benchmarking health IT among OECD countries: Better data for better policy. *Journal of the American Medical Informatics Association*, 21, 111–116.

Adolfsson, E.T., Rosenblad, A. and Wikblad, K. 2010. The Swedish National Survey of the quality and organization of diabetes care in primary healthcare – Swed-QOP. *Primary Care Diabetes*, 4, 91–97.

Agvall, B., Alehagen, U. and Dahlström, U. 2013. The benefits of using a heart failure management programme in Swedish primary healthcare. *European Journal of Heart Failure*, 15, 228–236.

Ahlén, G.C. and Gunnarsson, R.K. 2013. The physician's self-evaluation of the consultation and patient outcome: A longitudinal study. *Scandinavian Journal of Primary Health Care*, 31, 26–30.

Alenius, L.S., Tishelman, C., Runesdotter, S. and Lindqvist, R. 2014. Staffing and resource adequacy Strongly related to Rns' assessment of patient safety: A national study of Rns working in acute-care hospitals in Sweden. *BMJ Quality and Safety*, 23, 242–249.

Ammendrup, A., Bendixen, A., Sander, P., Ottesen, B. and Lose, G. 2009a. Urinary incontinence surgery in Denmark 2001–2003. *Ugeskrift for Laeger*, 171, 399–404.

Ammendrup, A., Jørgensen, A., Sander, P., Ottesen, B. and Lose, G. 2009b. A Danish national survey of women operated with mid-urethral slings in 2001. *Acta Obstetricia et Gynecologica Scandinavica*, 88, 1227–1233.

Ammentorp, J., Rørmann, D., Mainz, J. and Larsen, L.M. 2001. Measurement of the quality of care in a paediatric department. *Ugeskrift for Laeger*, 163, 7048–7052.

Andersson, A.C., Elg, M., Perseius, K.I. and Idvall, E. 2013. Evaluating a questionnaire to measure improvement initiatives in Swedish healthcare. *BMC Health Services Research*, 13, 47–56.

André, B., Sjøvold, E., Rannestad, T. and Ringdal, G.I. 2013. The impact of work culture on quality of care in nursing homes – A review study. *Scandinavian Journal of Caring Sciences* 28(3), 449–457.

Anell, A., Glenngård, A. and Merkur, S. 2012. Health system review. *Health Systems in Transition*, 14, 151–159.

Bakken, K., Larsen, E., Lindberg, P.C., Rygh, E. and Hjortdahl, P. 2007. Insufficient communication and information regarding patient medication in the primary healthcare. *Tidsskrift for Den Norske Lægeforening: Tidsskrift for Praktisk Medicin, Ny Række*, 127, 1766–1769.

Behr, L., Grit, K., Bal, R. and Robben, P. 2015. Framing and reframing critical incidents in hospitals. *Health, Risk & Society*, 17, 81–97.

Berg, L., Källberg, A.S., Göransson, K.E., Östergren, J., Florin, J. and Ehrenberg, A. 2013. Interruptions in emergency department work: An observational and interview study. *BMJ Quality and Safety*, 22, 656–663.

Berland, A., Gundersen, D. and Bentsen, S.B. 2012. Patient safety and falls: A qualitative study of home care nurses in Norway. *Nursing and Health Sciences*, 14, 452–457.

Bernstein, K., Bruun-Rasmussen, M. and Vingtoft, S. 2006. A method for specification of structured clinical content in electronic health records. *Studies in Health Technology and Informatics*, 124, 515–521.

Björkenstam, E., Ljung, R., Burström, B., Mittendorfer-Rutz, E., Hallqvist, J. and Weitoft, G.R. 2012. Quality of medical care and excess mortality in psychiatric patients – A nationwide register-based study in Sweden. *BMJ Open*, 2, e000778.

Bjorkman, M. and Malterud, K. 2009. Lesbian women's experiences with health care: A qualitative study. *Scandinavian Journal of Primary Health Care*, 27, 238–243.

Bjørn, B., Anhøj, J. and Lilja, B. 2009. Reporting of patient safety incidents: Experience from five years with a national reporting system. *Ugeskrift for Laeger*, 171, 1677–1680.

Bjørnstad, C.C.L., Gjertsen, E., Thorup, F., Gundersen, T., Tobiasson, K. and Otterstad, J.E. 2012. Temporary cardiac pacemaker treatment in five Norwegian regional hospitals. *Scandinavian Cardiovascular Journal*, 46, 137–143.

Brännström, M., Hägglund, L., Fürst, C.J. and Boman, K. 2012. Unequal care for dying patients in Sweden: A comparative registry study of deaths from heart disease and cancer. *European Journal of Cardiovascular Nursing: Journal of the Working Group on Cardiovascular Nursing of the European Society of Cardiology*, 11, 454–459.

Brattheim, B., Faxvaag, A. and Seim, A. 2011. Process support for risk mitigation: A case study of variability and resilience in vascular surgery. *BMJ Quality & Safety*, 20, 672–679.

Browall, M., Koinberg, I., Falk, H. and Wijk, H. 2013. Patients' experience of important factors in the healthcare environment in oncology care. *International Journal of Qualitative Studies on Health & Well-Being*, 8, 1–10.

Burns, N., Grove, S. and Gray, J. 2011. *Understanding Nursing Research*. Maryland Heights, MO: Elsevier Saunders.

Bylund, C. 2008. Structured patient-clinician communication using DIALOG improves patient quality of life. *Evidence Based Mental Health*, 11, 89.

Bystedt, M., Eriksson, M. and Wilde-Larsson, B. 2011. Delegation within municipal health care. *Journal of Nursing Management*, 19, 534–541.

Coulter, A. 2011. *Engaging Patient in Healthcare*. Berkshire, England: Open University Press.

Donabedian, A. 2005. Evaluating the quality of medical care. *Milbank Quarterly*, 83, 691–729.

Eriksson, H., Bergbrant, I.M., Berrum, I. and Mörck, B. 2011. Reducing queues: Demand and capacity variations. *International Journal of Health Care Quality Assurance*, 24, 592–600.

Fossum, M., Ehnfors, M., Svensson, E., Hansen, L.M. and Ehrenberg, A. 2013. Effects of a computerized decision support system on care planning for pressure ulcers and malnutrition in nursing homes: An intervention study. *International Journal of Medical Informatics*, 82(10), 911–921.

Fossum, M., Söderhamn, O., Cliffordson, C. and Söderhamn, U. 2012. Translation and testing of the Risk Assessment Pressure Ulcer Sore scale used among residents in Norwegian nursing homes. *BMJ Open*, 2(5), e001575.

Francis, R. 2013. Report of the mid Staffordshire NHS Foundation Trust Public Inquiry: Executive summary. London, U.K.: The Stationery Office: Mid Staffordshire NHS Foundation Trust Public Inquiry.

Frank, C., Asp, M., Fridlund, B. and Baigi, A. 2011. Questionnaire for patient participation in emergency departments: Development and psychometric testing. *Journal of Advanced Nursing*, 67, 643–651.

Frølich, A., Høst, D., Schnor, H., Nørgaard, A., Ravn-Jensen, C., Borg, E. and Hendriksen, C. 2010. Integration of healthcare rehabilitation in chronic conditions. *International Journal of Integrated Care*, 10, e033.

Gunningberg, L., Fogelberg-Dahm, M. and Ehrenberg, A. 2009. Improved quality and comprehensiveness in nursing documentation of pressure ulcers after implementing an electronic health record in hospital care. *Journal of Clinical Nursing*, 18, 1557–1564.

Häggström, M., Asplund, K. and Kristiansen, L. 2014. Important quality aspects in the transfer process. *International Journal of Health Care Quality Assurance*, 27, 123–139.

Hajdu, A., Eriksen, H.M., Sorknes, N.K., Hauge, S.H., Loewer, H.L., Iversen, B.G. and Aavitsland, P. 2011. Evaluation of the national surveillance system for point-prevalence of healthcare-associated infections in hospitals and in long-term care facilities for elderly in Norway, 2002–2008. *BMC Public Health*, 11, 923.

Hakkarainen, K.M., Andersson Sundell, K., Petzold, M. and Hägg, S. 2013. Prevalence and perceived preventability of self-reported adverse drug events – A population-based survey of 7099 adults. *PLoS ONE*, 8, e73166.

Håkonsen, H., Hopen, H.S., Abelsen, L., Ek, B. and Toverud, E.L. 2010. Generic substitution: A potential risk factor for medication errors in hospitals. *Advances in Therapy*, 27, 118–126.

Härkänen, M., Turunen, H., Saano, S. and Vehviläinen-Julkunen, K. 2013. Medication errors: What hospital reports reveal about staff views. *Nursing Management*, 19, 32–37.

Hedsköld, M., Pukk-Härenstam, K., Berg, E., Lindh, M., Soop, M., Øvretveit, J. and Sachs, M.A. 2013. Psychometric properties of the hospital survey on patient safety culture, HSOPSC, applied on a large Swedish health care sample. *BMC Health Services Research*, 13, 332.

Helmiö, P., Takala, A., Aaltonen, L.M. and Blomgren, K. 2012. WHO Surgical Safety Checklist in otorhinolaryngology – Head and neck surgery: Specialty-related aspects of check items. *Acta Oto-Laryngologica*, 132, 1334–1341.

Hofoss, D. and Deilkås, E. 2008. Roadmap for patient safety research: Approaches and roadforks. *Scandinavian Journal of Public Health*, 36, 812–817.

Hollman, D., Lennartsson, S. and Rosengren, K. 2014. District nurses' experiences with the free-choice system in Swedish primary care. *British Journal of Community Nursing*, 19, 30–35.

Holmboe, O., Bjerkan, A.M. and Skudal, K.E. Pasienterfaringer med norske sykehus i 2013. Resultater på sykehusnivå. PasOpp-rapport nr. 3 - 2014–2014.

Hovlid, E. and Bukve, O. 2014. A qualitative study of contextual factors' impact on measures to reduce surgery cancellations. *BMC Health Services Research*, 14, 215.

Hovlid, E., Bukve, O., Haug, K., Aslaksen, A.B. and Von Plessen, C. 2012. Sustainability of healthcare improvement: What can we learn from learning theory? *BMC Health Services Research*, 12, 235.

Ingemansson, M., Wettermark, B., Jonsson, E.W., Bredgård, M., Jonsson, M., Hedlin, G. and Kiessling, A. 2012. Adherence to guidelines for drug treatment of asthma in children: Potential for improvement in Swedish primary care. *Quality in Primary Care*, 20, 131–139.

IOM 2001. *Crossing the Quality Chasm: A New Health System for the 21st Century.* Washington, DC: National Academies Press.

Itoh, K., Andersen, H.B., Madsen, M.D., Østergaard, D. and Ikeno, M. 2006. Patient views of adverse events: Comparisons of self-reported healthcare staff attitudes with disclosure of accident information. *Applied Ergonomics*, 37, 513–523.

Jakobsen, D.H., Rud, K., Kehlet, H. and Egerod, I. 2014. Standardising fast-track surgical nursing care in Denmark. *British Journal of Nursing*, 23, 471–476.

Jakobsson, J. and Wann-Hansson, C. 2013. Nurses' perceptions of working according to standardized care plans: A questionnaire study. *Scandinavian Journal of Caring Sciences*, 27, 945–952.

Jensen, N.K., Nielsen, S.S. and Krasnik, A. 2010. Expert opinion on "best practices" in the delivery of health care services to immigrants in Denmark. *Danish Medical Bulletin*, 57, A4170.

Jha, A.K., Prasopa-Plaizier, N., Larizgoitia, I. and Bates, D.W. 2010. Patient safety research: An overview of the global evidence. *Quality & Safety in Health Care*, 19, 42–47.

Johansen, B., Mainz, J., Sabroe, S., Manniche, C. and Leboeuf-Yde, C. 2004. Quality improvement in an outpatient department for subacute low back pain patients: Prospective surveillance by outcome and performance measures in a health technology assessment perspective. *Spine*, 29, 925–931.

Johansen, I. and Rasmussen, M. 2009. Electronic requests for laboratory tests by general practitioners greatly reduce errors and costs. *Journal on Information Technology in Healthcare*, 7, 49–57.

Johansson, A., Andershed, B. and Anderzen-Carlsson, A. 2014. Conceptions of mental health care – From the perspective of parents' of adult children suffering from mental illness. *Scandinavian Journal of Caring Sciences*, 28(3), 496–504.

Johansson, M., Hägg, S. and Wallerstedt, S. 2011. Impact of information letters on the reporting rate of adverse drug reactions and the quality of the reports: A randomized controlled study. *BMC Clinical Pharmacology*, 11, 14.

Johnson, J.K., Arora, V.M. and Barach, P.R. 2013. What can artefact analysis tell us about patient transitions between the hospital and primary care? Lessons from the HANDOVER project. *European Journal of General Practice*, 19, 185–193.

Kaakinen, P., Kyngäs, H. and Kääriäinen, M. 2013. Predictors of good-quality counselling from the perspective of hospitalised chronically ill adults. *Journal of Clinical Nursing*, 22, 2704–2713.

Källman, U. and Suserud, B.O. 2009. Knowledge, attitudes and practice among nursing staff concerning pressure ulcer prevention and treatment – A survey in a Swedish healthcare setting. *Scandinavian Journal of Caring Sciences*, 23, 334–341.

Kennedy, I. 2001. The report of the public inquiry into children's heart surgery at the Bristol Royal Infirmary 1984–1995: Learning from Bristol. BRI Inquiry Final Report. Retrieved August 15, 2015 http://webarchive.nationalarchives.gov.uk/20090811143745/http:/www.bristol-inquiry.org.uk/final_report/the_report.pdf.

Khorram-Manesh, A., Hedelin, A. and Ortenwall, P. 2009. Hospital-related incidents; causes and its impact on disaster preparedness and prehospital organisations. *Scandinavian Journal of Trauma, Resuscitation and Emergency Medicine*, 17, 26.

Kivekäs, E., Luukkonen, I., Mykkänen, J. and Saranto, K. 2014. Improving the coordination of patients' medication management: A regional Finnish development project. *Studies in Health Technology and Informatics*, 201, 175–180.

Kohn, L.T., Corrigan, J. and Donaldson, M.S. 2000. *To Err Is Human: Building a Safer Health System*. Washington, DC: National Academy Press.

Kristoffersen, A.H., Thue, G. and Sandberg, S. 2006. Postanalytical external quality assessment of warfarin monitoring in primary healthcare. *Clinical Chemistry*, 52, 1871–1878.

Kuusela, M., Koivisto, A.L., Vainiomäki, P., Vahlberg, T. and Rautava, P. 2011. The medico-professional quality of GP consultations assessed by analysing patient records. *Scandinavian Journal of Primary Health Care*, 29, 222–226.

Lefebvre, C., Manheimer, E., and Glanville, J. Chapter 6: Searching for studies. In: Higgins, J.P.T., Green, S. (eds.), *Cochrane Handbook for Systematic Reviews of Interventions Version* 5.1.0 (updated March 2011). The Cochrane Collaboration, 2011. Available from www.cochrane-handbook.org.

Lipczak, H., Knudsen, J.L. and Nissen, A. 2011. Safety hazards in cancer care: Findings using three different methods. *BMJ Quality and Safety*, 20, 1052–1056.

Martin, G. and Dixon-Woods, M. 2014. After Mid Staffordshire: From acknowledgement, through learning, to improvement. *BMJ Quality & Safety*, 23, 706–708.

Mattsson, T.O., Knudsen, J.L., Lauritsen, J., Brixen, K. and Herrstedt, J. 2013. Assessment of the global trigger tool to measure, monitor and evaluate patient safety in cancer patients: Reliability concerns are raised. *BMJ Quality and Safety*, 22, 571–579.

NORDEN. 2013. Nordisk Samarbejde på social- og sundhedsomårdet. Strategi for social og sundhedsområdet fra 2013 og frem. Copenhagen, Denmark: Nordisk Ministerråd.

NORDEN. 2014. Does the Nordic Model need to change?: Overview of the research report, The Nordic model – Challenged but capable of reform. Copenhagen, Denmark: Nordic Council of Ministers.

Nørgaard, B., Kofoed, P.E., Ohm Kyvik, K. and Ammentorp, J. 2012. Communication skills training for health care professionals improves the adult orthopaedic patient's experience of quality of care. *Scandinavian Journal of Caring Sciences*, 26, 698–704.

OECD. 2013. *Health at a Glance 2013: OECD Indicators*, OECD Publishing. http://dx.doi.org/10.1787/health_glance-2013-en.

Oksuzyan, A., Jeune, B., Juel, K., Vaupel, J.W. and Christensen, K. 2013. Changes in hospitalisation and surgical procedures among the oldest-old: A follow-up study of the entire Danish 1895 and 1905 cohorts from ages 85 to 99 years. *Age & Ageing*, 42, 476–481.

Olsen, E. and Aase, K. 2010. A comparative study of safety climate differences in healthcare and the petroleum industry. *Quality & Safety in Health Care*, 19(Suppl. 3), i75–i79.

Oltedal, S., Garratt, A., Bjertnæs, Ø., Bjørnsdottìr, M., Freil, M. and Sachs, M. 2007. The NORPEQ patient experiences questionnaire: Data quality, internal consistency and validity following a Norwegian inpatient survey. *Scandinavian Journal of Public Health*, 35, 540–547.

Pearson, A. and Aromataris, E. 2009. Patient Safety in Primary Healthcare: A review of the literature. Adelaide, Australia: Australian Commission on Safety and Quality in Health Care, The Joanna Briggs Institute.

Pedersen, R., Nortvedt, P., Nordhaug, M., Slettebø, Å., Grøthe, K.H., Kirkevold, M., Brinchmann, B.S. and Andersen, B. 2008. In quest of justice? Clinical prioritisation in healthcare for the aged. *Journal of Medical Ethics*, 34, 230–235.

Petersson, I.F., Strombeck, B., Andersen, L., Cimmino, M., Greiff, R., Loza, E., Scire, C. et al. 2014. Development of healthcare quality indicators for rheumatoid arthritis in Europe: The eumusc.net project. *Annals of the Rheumatic Disease*, 73, 906–908.

Pronovost, P., Goeschel, C., Marsteller, J., Sexton, J., Pham, J. and Berenholtz, S. 2009. Framework for patient safety research and improvement. *Circulation*, 119, 330–337.

Putzer, G., Braun, P., Zimmermann, A., Pedross, F., Strapazzon, G., Brugger, H. and Paal, P. 2013. LUCAS compared to manual cardiopulmonary resuscitation is more effective during helicopter rescue – A prospective, randomized, cross-over manikin study. *The American Journal of Emergency Medicine*, 31, 384–389.

Rabøl, L.I., Andersen, M.L., Østergaard, D., Bjørn, B., Lilja, B. and Mogensen, T. 2011. Descriptions of verbal communication errors between staff. An analysis of 84 root cause analysis – Reports from Danish hospitals. *BMJ Quality & Safety*, 20, 268–274.

Rahimi, B., Moberg, A., Timpka, T. and Vimarlund, V. 2008. Implementing an integrated computerized patient record system: Towards an evidence-based information system implementation practice in healthcare. In *AMIA 2008 Annual Symposium, Biomedical and Health Informatics: Form Foundations to Applications to Policy*, Washington DC, 8–12 November, 2008, pp. 616–620.

Rahimi, B., Timpka, T., Vimarlund, V., Uppugunduri, S. and Svensson, M. 2009. Organization-wide adoption of computerized provider order entry systems: A study based on diffusion of innovations theory. *BMC Medical Informatics and Decision Making*, 9, 52.

Ranji, A., Dykes, A.-K. and Ny, P. 2012. Routine ultrasound investigations in the second trimester of pregnancy: The experiences of immigrant parents in Sweden. *Journal of Reproductive & Infant Psychology*, 30, 312–325.

Röing, M., Rosenqvist, U. and Holmström, I.K. 2013. Threats to patient safety in telenursing as revealed in Swedish telenurses' reflections on their dialogues. *Scandinavian Journal of Caring Sciences*, 27, 969–976.

Rytter, L., Jakobsen, H.N., Rønholt, F., Hammer, A.V., Andreasen, A.H., Nissen, A. and Kjellberg, J. 2010. Comprehensive discharge follow-up in patient's homes by GPs and district nurses of elderly patients. *Scandinavian Journal of Primary Health Care*, 28, 146–153.

Sampo, M.M., Rönty, M., Tarkkanen, M., Tukiainen, E.J., Böhling, T.O. and Blomqvist, C.P. 2012. Soft tissue sarcoma – A population-based, nationwide study with special emphasis on local control. *Acta Oncologica*, 51, 706–712.

Schildmeijer, K., Nilsson, L., Arestedt, K. and Perk, J. 2012. Assessment of adverse events in medical care: Lack of consistency between experienced teams using the global trigger tool. *BMJ Quality & Safety*, 21, 307–314.

Schildmeijer, K., Unbeck, M., Muren, O., Perk, J., Pukk Härenstam, K. and Nilsson, L. 2013. Retrospective record review in proactive patient safety work – Identification of no-harm incidents. *BMC Health Services Research*, 21, 307–314.

Schmidt, S., Thyen, U., Chaplin, J., Mueller-Godeffroy, E. and Bullinger, M. 2008. Healthcare needs and healthcare satisfaction from the perspective of parents of children with chronic conditions: The DISABKIDS approach towards instrument development. *Child: Care, Health and Development*, 34, 355–366.

Siersma, V., Waldemar, G. and Waldorff, F.B. 2013. Subjective memory complaints in primary care patients and death from all causes: A four-year follow-up. *Scandinavian Journal of Primary Health Care*, 31, 7–12.

Sigurgeirsdóttir, S., Waagfjörð, J. and Maresso, A. 2014. Iceland: Health system review. *Health Systems in Transition*, 16, 1.

Sinnemäki, J., Sihvo, S., Isojärvi, J., Blom, M., Airaksinen, M. and Mäntylä, A. 2013. Automated dose dispensing service for primary healthcare patients: A systematic review. *Systematic Reviews*, 2, 1.

Sipilä, R., Ketola, E., Tala, T. and Kumpusalo, E. 2008. Facilitating as a guidelines implementation tool to target resources for high risk patients – The Helsinki Prevention Programme (HPP). *Journal of Interprofessional Care*, 22, 31–44.

Sjøhart Lund, M.L., Kamarauskas, A.G., Mainz, J. and Ovesen, T. 2010. Quality of outpatient tonsillectomy performed in ear, nose & throat practice. *Ugeskrift for Laeger*, 172, 2049–2054.

Skarstein, J., Dahl, A.A., Laading, J. and Fosså, S.D. 2002. 'Patient satisfaction' in hospitalized cancer patients. *Acta Oncologica*, 41, 639–645.

Skudal, K.E., Garratt, A.M., Eriksson, B., Leinonen, T., Simonsen, J. and Bjertnaes, O.A. 2012. The Nordic Patient Experiences Questionnaire (NORPEQ): Cross-national comparison of data quality, internal consistency and validity in four Nordic countries. *BMJ Open*, 2, e000864.

Smith, F., Carlsson, E., Kokkinakis, D., Forsberg, M., Kodeda, K., Sawatzky, R., Friberg, F. and Öhlén, J. 2014. Readability, suitability and comprehensibility in patient education materials for Swedish patients with colorectal cancer undergoing elective surgery: A mixed method design. *Patient Education and Counseling*, 94, 202–209.

Solheim, O., Jakola, A.S., Gulati, S. and Johannesen, T.B. 2012. Incidence and causes of perioperative mortality after primary surgery for intracranial tumors: A national, population-based study. *Journal of Neurosurgery*, 116, 825–834.

Steinø, P., Jørgensen, C.B. and Christoffersen, J.K. 2013. Psychiatric claims to the Danish Patient Insurance Association have low recognition percentages. *Danish Medical Journal*, 60, A4621.

Stepnick, L. and Findlay, S. 2003. *Accelerating Quality Improvement in Health Care: Strategies to Speed the Diffusion of Evidence-Based Innovations*. Washington, DC: National Institute for Health Care Management Foundation.

Struwe, J., Dumpis, U., Gulbinovic, J., Lagergren, A. and Bergman, U. 2006. Healthcare associated infections in university hospitals in Latvia, Lithuania and Sweden: A simple protocol for quality assessment. *Euro Surveillance*, 11, 167–171.

Thor, J., Herrlin, B., Wittlöv, K., Øvretveit, J. and Brommels, M. 2010. Evolution and outcomes of a quality improvement program. *International Journal of Health Care Quality Assurance*, 23, 312–327.

Thorstad, M., Sie, I. and Andersen, B.M. 2011. MRSA: A challenge to Norwegian nursing home personnel. *Interdisciplinary Perspectives on Infectious Diseases*, 2011, 197683.

Unbeck, M., Dalen, N., Muren, O., Lillkrona, U. and Härenstam, K.P. 2010. Healthcare processes must be improved to reduce the occurrence of orthopaedic adverse events. *Scandinavian Journal of Caring Sciences*, 24, 671–677.

Unbeck, M., Muren, O. and Lillkrona, U. 2008. Identification of adverse events at an orthopedics department in Sweden. *Acta Orthopaedica*, 79, 396–403.

Vægter, K., Wahlström, R. and Svärdsudd, K. 2012. General practitioners' awareness of their own drug prescribing profiles after postal feedback and outreach visits. *Upsala Journal of Medical Sciences*, 117, 439–444.

Wallin, C.J. and Thor, J. 2008. SBAR – Modell för bättre kommunikation mellan vårdpersonal: Ineffektiv kommunikation bidrar till majoriteten av skador i vården. *Lakartidningen*, 105, 1922–1925.

Weick, K.E. and Sutcliffe, K.M. 2003. Hospitals as cultures of entrapment: A re-analysis of the Bristol Royal Infirmary. *California Management Review*, 45, 73–84.

Wessel, M., Lynøe, N., Juth, N. and Helgesson, G. 2012. The tip of an iceberg? A cross-sectional study of the general public's experiences of reporting healthcare complaints in Stockholm, Sweden. *BMJ Open*, 2, e000489.

Wettermark, B., Bergman, U. and Krakau, I. 2006a. Using aggregate data on dispensed drugs to evaluate the quality of prescribing in urban primary health care in Sweden. *Public Health*, 120, 451–461.

Wettermark, B., Persson, M., Wilking, N., Kalin, M., Korkmaz, S., Hjemdahl, P., Godman, B., Petzold, M. and Gustafsson, L. 2010. Forecasting drug utilization and expenditure in a metropolitan health region. *BMC Health Services Research*, 10, 128–128.

Wettermark, B., Tomson, G. and Bergman, U.L.F. 2006b. Kvalitetsindikatorer för läke-medel – Läget i Sverige idag. *Lakartidningen*, 103, 3607–3611.

Whittemore, R. and Knafl, K. 2005. The integrative review: Updated methodology. *Journal of Advanced Nursing*, 52, 546–553.

Wiig, S., Storm, M., Aase, K., Gjestsen, M.T., Solheim, M., Harthug, S., Robert, G. and Fulop, N. 2013. Investigating the use of patient involvement and patient experience in quality improvement in Norway: Rhetoric or reality? *BMC Health Services Research*, 13, 206.

Wolf, A., Olsson, L.-E., Taft, C., Swedberg, K. and Ekman, I. 2012. Impacts of patient characteristics on hospital care experience in 34,000 Swedish patients. *BMC Nursing*, 11, 8–8.

Öhrn, A. and Eriksson, G. 2007. Risk analysis – A tool for IT development and patient safety a comparative study of weaknesses before and after implementation of a health care system in the county Council of Ostergotland, Sweden. In: Kuhn, K.A., Warren, J.R., and Leong, T.-Y. (eds.), *Medinfo 2007: Proceedings of the 12th World Congress on Health (Medical) Informatics; Building Sustainable Health Systems*. Amsterdam: IOS Press, 2007, pp. 2359–2360. Studies in health technology and informatics, ISSN 0926-9630; v. 129.

Øvretveit, J. 2002. Producing useful research about quality improvement. *International Journal of Health Care Quality Assurance*, 15, 294–302.

Section I

Perspectives on Patient Safety and Quality in Healthcare

2 Patient-Centred Care in the Nordic Countries

Marianne Storm and Angela Coulter

CONTENTS

This chapter looks at the quality of healthcare through patients' eyes, focusing on policies and practices in the Nordic countries (Norway, Denmark, Sweden and Finland) aimed at ensuring that health services are responsive to the needs of the people who use them. We describe some key concepts and outline the challenges faced by those wanting to move health systems in a more patient-centred direction.

QUALITY THROUGH PATIENTS' EYES

Paternalistic care models, based on assumptions that healthcare professionals know what is in the best interest of their patients and can make decisions on their behalf without involving them, are now being challenged (Coulter, 2011). Over the last 20 years, there has been a shift from viewing patients as passive recipients of healthcare to considering them as active participants. This demands new types of relationships between patients and health professionals that emphasise collaboration, information sharing and involvement in treatment decisions (Ocloo and Fulop, 2012). The patient's perspective is seen as a valuable source of knowledge about health and disease and measuring their experiences and acting to tackle any problems identified as a central aim of quality improvement programmes. The goal is to ensure that services are responsive to patients' experiences, needs and values (Institute of Medicine, 2001).

Patient-centredness is an overarching aim for improving the quality of healthcare in many countries (Docteur and Coulter, 2012). The U.S. Institute of Medicine (IOM) defines patient-centredness as care that is respectful and responsive to individuals' preferences, needs and values and ensures that patients' own values guide clinical decisions (IOM, 2001). The IOM report emphasises the importance of shared knowledge between professionals and patients, patients' rights to see their own medical information and clinical records, transparency and access to information on system performance for patients and their families. Co-ordination, continuity and effective team work are key attributes of a patient-centred system. Opportunities for collective engagement and involvement in health service development are also considered important for improving service quality (Andreassen, 2005; Coulter, 2011; Crawford et al., 2003).

In reviewing the literature on patient-centred care, Mead and Bower (2000) pointed to several distinctive elements: the incorporation of social and psychological factors for understanding diseases and treatment effects; healthcare professionals who see the person in the patient and are aware of the patient's experiences and feelings and shared responsibility between health professionals and patients, including shared decision-making. These factors form the basis of a therapeutic alliance where patients and professionals are expected to reach agreement on treatment goals and management plans. There is encouraging evidence that education and training of health professionals in patient-centred treatment and care can lead to improved communications, greater satisfaction and more appropriate decision-making (Lewin et al., 2009). Evidence shows that shared decision-making helps to increase patients' knowledge, leading to greater confidence in their choices and more active participation in treatment (Duncan et al., 2010; Storm and Edwards, 2013). Reported barriers to implementation of shared decision-making include perceptions of lack of time and lack of fit with traditional clinical routines. Motivating health professionals is seen as one key to success (Légaré et al., 2010).

High-quality care means acknowledging the perspectives of service users, their experiences of and expectations of service provision (Andreassen, 2005). User involvement goes beyond the individual patient's influence on his or her own treatment and care to encompass influencing the development and delivery of healthcare services (Andreassen, 2005; Crawford et al., 2003). For example, Crawford et al. (2003) list numerous methods adopted by mental health trusts in London to enhance service users' influence on service development: users are involved in staff appointments; they contribute to staff training programmes; they help to assess the quality of services and they are involved in the governance of the organisations, attending trust board meetings. They are also actively involved in designing, delivering and disseminating health research, particularly in studies that aim to improve safety and quality (Beresford, 2005; Smith et al., 2009).

Tritter (2009) proposed a framework for user involvement in health services encompassing multiple objectives. His framework differentiates between five categories of involvement:

1. Decisions about treatment and care
2. Service development
3. Service evaluation
4. Education and training of health professionals
5. Research

He draws a distinction between individual and collective involvement, contrasting situations when patients and users act on their own behalf to acting on behalf of groups or populations. He also distinguishes between direct involvement where service users have the power and authority to participate in decisions and indirect involvement where users contribute information, but the decisions are made by health professionals. The framework encompasses both reactive user involvement, where patients respond to a predefined agenda, and proactive involvement, where service users shape the agenda and set their own objectives. Direct proactive involvement is a key feature of experienced based co-design where patient and professionals participate in a face-to-face collaborative venture to co-design services (Bate and Robert, 2006).

THE NORDIC HEALTHCARE SETTING

Norway, Denmark, Sweden and Finland are parliamentary democracies operating at three administrative levels: the state, the regions and the municipalities. The common view in these Nordic countries is that citizens in a democracy should not only be governed but also have a right and a responsibility to take part in governing. Civil rights are protected, ensuring that the law applies equally to all members of the community, including the right to vote and to access health and welfare (Eriksen and Weigård, 1993). The welfare states in the Nordic countries are underpinned by the principles of equal opportunities, social solidarity and security for all people regardless of characteristics such as gender, age, ethnicity, sexual orientation or religion (Eriksen and Weigård, 1993; Nordisk Ministerråd, 2013).

We will now describe the ways in which Nordic policymakers are attempting to move healthcare systems towards greater patient-centredness by involving patient organisations; promoting patient rights and encouraging patient choice, patient information and patient feedback.

POLICIES

Improving quality and safety in the health, welfare and social services is a political priority for the Nordic co-operation on social affairs and health (Nordisk Ministerråd, 2013). The main emphasis is on ensuring that health, care and welfare services are effective, safe, co-ordinated, involving service users and being responsive to their needs.

Strengthening the patients' role in healthcare services has been a government priority in Norway for two decades (Norwegian Directorate of Health, 2005; Norwegian Ministry of Health and Care Services, 1996). In 2005, the Norwegian Directorate of Health launched a strategy for quality improvement for the period 2005–2015. In common with the IOM in the United States (IOM, 2001), the strategy defines healthcare quality as that which is effective, safe, patient-centred, co-ordinated, resource effective and accessible (Deilkås et al., 2015; Norwegian Directorate of Health, 2005). Norway launched its first national patient safety campaign in 2011, and the first report to Parliament on quality and safety in healthcare was issued in 2012.

In Denmark, the Health and Medicine Authority published the first National Strategy for Quality in 1993. The strategy addressed the responsibilities of leaders and health professionals, and it was followed by a specific strategy on national quality registries. National surveys on patients' experiences have been carried out annually since 2000. Patient-centred care has only recently become an explicit goal of Danish healthcare (Knudsen et al., 2015). In April 2015, the Danish Ministry of Health launched the National Quality Programme for Health 2015–2018 with a particular emphasis on patient-centred care and involvement of patients and their next of kin and training of professionals to facilitate a patient-centred approach (Danish Ministry of Health, 2015).

In Sweden, reducing waiting time, increasing efficiency and patient choice have been key priorities in national healthcare reforms (Øvretveit et al., 2015). In 2010, a law was passed giving patients a choice of primary healthcare centre, and in 2011, an Act setting out healthcare providers responsibilities for patient safety was introduced (Øvretveit et al., 2015).

In January 2008, the Finnish government adopted a National Development Programme for Social Welfare and Health Care. The central themes of the programme were to reduce inequalities in health and welfare and increase the quality, effectiveness and accessibility of health and social services. Strengthening the position of users (patient empowerment) was an important goal (Vuorenkoski et al., 2008).

PATIENT ORGANISATIONS

Engaging citizens in health is considered important to increase public understanding of health issues, to reduce health inequalities, to encourage democratic accountability and to ensure that the health system is responsive to people's needs and preferences (Coulter, 2011). Voluntary patient organisations are important channels for public engagement. In each of the Nordic countries, there are a large number of patient organisations, commonly formed around particular diseases or health problems. The organisations vary in size and in the extent to which they are professionally run (Olejaz et al., 2012; Ringard et al., 2013; Winblad and Ringard, 2009). They provide information, help and support to patients; they maintain dialogues with the relevant health authorities, engage in service delivery and support research (Anell et al., 2012; Olejaz et al., 2012; Toiviainen et al., 2010; Vuorenkoski et al., 2008). The larger patient organisations in the Nordic countries aim to influence health policy in various ways. They participate in parliamentary hearings and act as members of publicly appointed boards and councils (Olejaz et al., 2012; Winblad and Ringard, 2009).

In Norway, patient groups have played an important role in setting the health policy agenda ensuring that patients have the right to information about their health status and the ability to participate in decisions about their care and treatment (Andreassen, 2005). The Norwegian Federation of Organisations of Disabled People (FFO) is an umbrella organisation with 80 member organisations for people with disabilities and chronic diseases representing about 335,000 members. Their main goal is to improve living conditions and the fulfilment of human rights for all those with disabilities and chronic diseases (FFO, 2016).

There are between 200 and 300 patient groups in Denmark. Many of these groups have explicitly taken on policy advocacy, giving input into health debates and ensuring that patients' views are not neglected (Olejaz et al., 2012). Danish Patients is an umbrella organisation for 79 patient associations representing 870,000 members. They aim to contribute to the development of a patient-focused healthcare system through targeted efforts to ensure that patient involvement remains high on the political agenda (Danish Patients, 2015).

According to Anell et al. (2012) there are more than 100 patient and consumer organisations representing different patient groups in Sweden. The largest such organisation is the Swedish Rheumatism Association with about 50,000 members. Their primary goal is to support their members in their everyday efforts to cope with their condition, educating policymakers, for example, on the importance of access to rehabilitation and shorter waiting times for treatment (Swedish Rheumatism Association, 2016).

In Finland, there are about 130 patient organisations, some of which are members of national health and social welfare umbrella organisations (Toiviainen et al., 2010). The Cancer Society of Finland is one of the largest umbrella groups with 12 provincial cancer associations and 6 national patient organisations among its members (Cancer Society of Finland, 2016).

PATIENT RIGHTS

Health legislation complements and reinforces policies ensuring people equal access to health and social services. Norway, Finland and Denmark have established separate bills on patients' rights. Common themes addressed include the right to make informed decisions, the right to comprehensible information and decisions made in partnership between clinician and patient (Danish Health Act, 2014; Norwegian Patient Rights Act, 1999; The Finnish Act on the Status and Rights of Patients No. 785/1992). Information is necessary to access and benefit from treatment, and its importance is recognised as a prerequisite for making informed choices about treatment and to consent to treatment and healthcare (Kjellevold, 2005).

In Norway, the Patient Rights Act, which came into force in 2001, ensures the right to access healthcare, to receive information and to access personal health records and a right to confidentiality, to consent to or decline treatment, to participate in treatment decisions and to choose a hospital for treatment. Patients are entitled to information about their rights, duties and practices and to professional advice relevant to their individual needs. The right to participate in the implementation of healthcare is stated in the Patient Rights Act § 3-1. Participation is expected to be adapted to the individual's capacity. The Patients' Rights Act also entitles those with chronic or long-term conditions to an individual care plan. In Norway, patients also have the right to have medical decisions reviewed and, if necessary, reversed (Ringard et al., 2013).

In Denmark, an Act on Patient Rights was passed in 1998 focusing on rights to information, to self-determination, to informed consent and to ensuring that healthcare personnel respect patients' dignity, integrity and autonomy. This includes the right to access information about their health condition, test results, treatment options and risks of complications, adapted to the individuals' capacities and needs. Rights to share information with family members, to access their personal health records and to confidentiality are also enshrined in a revised version of the Act. This commits to easy and equal access to healthcare for all citizens in Denmark, to choice of health provider and hospital, together with a waiting-time guarantee to ensure prompt access to health services. Stated objectives also include general requirements regarding high quality of care, continuity of care (coherent and linked services) and transparency of the health system (ensuring quality indicators on clinical performance becoming available on the Internet) (Olejaz et al., 2012).

In Sweden, there is no specific law regulating patients' rights. Instead, Sweden uses multiple pieces of legislation to protect patients' rights, such as choice of provider and treatment, the right to information about health, the right to privacy and to access medical records and the right to a second opinion when suffering from a life-threatening disease (Anell et al., 2012). These are incorporated in other legislation and are formulated in policy agreements between the state and the county councils. In 2011, Sweden enacted a new patient safety law, which aims to protect everyone affected by healthcare, both patients and family members, giving new opportunities for influencing healthcare contents and report cases of wrong treatment (Øvretveit et al., 2015).

An act on patients' status and rights in Finland came into force in 1993 (the Act on the Status and Rights of Patients) (Vuorenkoski et al., 2008). It applies to every part of the healthcare system and to services provided in social welfare institutions. This act concerns the following issues: the right to appropriate healthcare and social services; the right to receive healthcare within the waiting-time limits; the right to information about health status, treatment and possible risks; the right to autonomy and informed consent to treatment; the right to confidentiality and to access relevant medical documents and the right to complain (The Finish National Supervisory Authority for Welfare and Health, 2012; Vuorenkoski et al., 2008).

PATIENT CHOICE

New public management (NPM) is an ideology that emerged following critiques of bureaucracy and inefficiency in the public sector (Stamsø, 2005). A core goal of NPM has been to increase the public sector's ability to deliver economically efficient services by adapting market models, principles and ideas from the private sector (Busch et al., 2003; Christensen and Lægreid, 2001; Glenngård et al., 2011). NPM emphasises that citizens should be viewed as consumers of public services, free to choose among different services and to participate in service planning (Busch et al., 2003). In healthcare, this approach represented a significant shift in focus from viewing patients as passive and dependent to seeing them as active and competent consumers. Patients are empowered to choose the services that fit their needs, preferences and are of high quality (Storm et al., 2009). An important component of this approach was the notion that people should be able to access information on the quality of care in different organisations to enable them to select the best (Coulter, 2011). It was hoped that this would provide both financial incentives and other types of 'market signals' to encourage providers to ensure that their services were of sufficiently high quality to attract customers (patients). There are several possible types of choices in healthcare in the Nordic countries, including choice of general practitioner (GP), hospital or type of provider (public, voluntary or private sector).

People in Norway can choose their GP. Patients also have the right to choose any public hospital across the country for elective care (Norwegian Patient Rights Act, 1999). The information service 'Helsenorge' supports patients' right to choose where to receive treatment and provides information on hospital waiting times for specific patient groups (Helsenorge, 2016). Norway has a system of individual waiting-time guarantees for patients (Ringard et al., 2013). According to the Norwegian Patient Rights Act (1999), all patients referred to a hospital have the right to have their health status evaluated within 30 working days. The hospital will decide whether there is need for medical care and provide information on when treatment is expected to take place. Patients with the same diagnosis may have different waiting times, depending on the severity of the disease and other factors specific to the individual (Norwegian Directorate of Health, 2006). Patients have an additional right to travel abroad if treatment cannot be provided in Norway or it cannot be provided within a given time frame (Ringard et al., 2013).

People in Denmark can register with a GP of their own choice, practising close to their home. Since 1993, patients have been free to choose any hospital in the country as long as the treatment takes place at the same level of specialisation (Olejaz et al., 2012). Denmark also has a waiting-time guarantee. If a patient is not offered treatment within 1 month of referral, he or she can choose treatment at any private hospital or clinic in the country by agreement with the health region or abroad (Olejaz et al., 2012).

Choice of primary care provider became mandatory in Sweden in January 2010 (The primary choice reform). Patients can register with any public or private provider/primary healthcare centre accredited by the local county council (Anell et al., 2012). Sweden has a waiting-time guarantee for elective treatment introduced in 2005, referred to as a '0-7-90-90' rule. This means instant contact with the primary health service system for a medical consultation, seeing a GP within 7 days, consulting specialist within 90 days referral and no more than 90 days to receive treatment following diagnosis (Anell et al., 2012; Øvretveit et al., 2015). If these criteria are not met, the county council must offer care at an alternative provider. There is a publicly available website that provides information about current waiting times (Anell et al., 2012; Väntetider i vården, 2016).

Since 2014, patients in Finland have had the right to choose any health centre or unit within their municipality, and any hospital providing specialist care, in consultation with the referring doctor. Finland has a maximum waiting-time limit for elective treatment. The need for specialist healthcare must be assessed within three weeks of referral, and further clinical assessments must commence within 3 months. Hospital treatment must begin within six months of the initial assessment (Finnish National Supervisory Authority for Welfare and Health, 2012; Vuorenkoski et al., 2008).

PATIENT INFORMATION

Improving access to and use of health information is essential for people to make informed choices about their health. People need relevant, reliable and comprehensible information provided in a timely manner. Printed or electronic information can improve patient's knowledge and their understanding of their condition, increase their sense of empowerment, improve coping ability and reduce anxiety (Coulter, 2011). In the Nordic countries, there are health information portals accessible via the Internet covering information on health and illness, patient rights and the quality of care to inform people's choice of provider. Each country has patient offices or patient ombudsmen providing information and advice on patient rights, choice and complaint procedures.

In Norway, the Internet portal 'helseNorge' disseminate information on patient rights and public health services (Helsenorge, 2016). Users can access information on fees, electronic prescriptions, vaccinations and how to change their GP. There is also information on health and disease to support self-care and involvement in the health system and on quality indicators to support hospital choice, with links to the Health and Social Services ombudsman. Every county in Norway has an ombudsman to safeguard patients' rights. The ombudsman provides advice and guidance on patients' rights, what to do if you experience an adverse event and assistance with formulating complaints (Ringard et al., 2013).

Every region in Denmark has a patient office and advice service, similar to a regional ombudsman (Winblad and Ringard, 2009). Their primary task is to provide patients with advice on their rights to choice of provider and treatment, how to make a complaint and claim compensation and waiting list guarantees. The patient advisor acts as a counsellor and a problem-solver in close dialogue between patients, relatives and hospital staff (Olejaz et al., 2012). In addition, Denmark has several websites established by the National Board on Health, Danish Regions and the Ministry of Health ensuring patients have access to information, including information about waiting times for operations and procedures and information about hospital performance and quality (Olejaz et al., 2012).

In Sweden, each county council and municipality has a patients' advisory committee to support individuals with information, promoting contact between patients, professionals and health services (Anell et al., 2012). There is also a health information portal which is a collaborative initiative by all county councils and regions in Sweden. The website provides written medical information and information about care pathways, a chat service and a 24-hour phone line where medical staff are available to give medical advice and advice on where to seek care if necessary (1177 Vårdguiden, 2016). Citizens may also create their own account in order to make healthcare appointments, renew drug prescriptions and obtain information about test results (Anell et al., 2012).

Finland has a website providing information about moving to and living in Finland, the Finish health service system and patient rights (Infopankki.fi, 2016). Individual patient organisations are also a valuable source of information. For example, the Cancer Society of Finland has a website providing information about types of cancer, patients' rights and a professional advice service available by e-mail, phone and chat (Cancer Society of Finland, 2016). All healthcare units in Finland must appoint a patient ombudsman obliged to inform patients about their rights and to assist with complaints about treatment (Winblad and Ringard, 2009).

PATIENT EXPERIENCE SURVEYS

Patient experience data are a key resource for assessing, monitoring and improving quality of care (Berwick, 2009; Coulter, 2011). Topics commonly covered in questionnaire surveys include access and waiting times, provision of information, communications with health or social care professionals, quality of the physical environment, involvement in decisions, support for self-care, co-ordination of care, quality of life and health status (Coulter et al., 2014). Patients' experience is positively associated with indicators of clinical effectiveness and patient safety, uptake of preventive care and appropriate resource use (Doyle et al., 2013). Isaac et al. (2010) reported an association between level of patient experience and several patient safety indicators. Better patient experiences were associated with reduced rates of pressure ulcers, fewer infections due to medical care and fewer post-operative complications such as bleeding, respiratory failure, pulmonary embolism and sepsis.

In Norway, Sweden and Denmark, national surveys of patients' experiences are conducted regularly, but this is not the case in Finland which has no national patient survey programme (Skudal et al., 2012). The Norwegian Knowledge Centre for the Health Services conducts national surveys of patients' experiences at Norwegian hospitals. These reveal both strengths and weaknesses; for example, hospital discharge was the area most complained about in 2011, 2012 and 2013 while patient safety achieved the highest scores, followed by experiences of patients' relatives and communications with nursing staff (Bjerkan et al., 2014). A study 'Patient safety and quality of health care in 2013' conducted with a random sample of the Norwegian population found that 42% of respondents had experienced an adverse event following contact with health services affecting themselves or a family member. Only 39% of those who had had this experience reported the event to staff inside or outside the hospital. Of those that did, most drew it to the attention of a doctor, nurse or pharmacist (41%), hospital manager (28%) or the Norwegian System for Compensation to Patients (26%) (Haugum et al., 2013).

The National Danish Survey of Patient Experiences is an annual nationwide questionnaire survey for assessing patients' experiences (Olejaz et al., 2012). Among all respondents in 2014, the most positive feature was perceptions of friendly staff, while reports on patient involvement, patient-perceived errors and provision of discharge information were more negative. Patients and their families complained about insufficient involvement in decisions about treatment and care, and many complained that they were not properly informed about the effects and side effects of new medications. Patients admitted via emergency settings reported the most negative experiences (Patientoplevelser, 2014).

A national population survey was initiated in Sweden in the 1990s to assess the general population's attitudes, knowledge and expectations towards healthcare performance and to compare responses in different county councils (Anell et al., 2012). In the 2014 survey, about 64% of respondents felt that waiting times at health centres were reasonable, and 77% knew that they could call 1177 for advice and assistance on healthcare (Befolkningsundersökning, 2014). Every 2 years, a National Swedish Patient Survey is conducted in all county councils and regions. The survey asks about communications with staff, information provision, participation, confidentiality, accessibility, perceived usefulness and overall satisfaction. In Spring 2014, the National Patient Survey was conducted among those experiencing inpatient specialised medical care. The most positive areas were perceived usefulness of treatment while the lowest scores related to involvement and accessibility (Swedish Patient Survey, 2014). A stated purpose of this survey is to guide people in their choice of health provider (Anell et al., 2012).

COMPLAINTS AND REDRESS

Patients in Norway, Sweden, Finland and Denmark have a right to complain about their medical treatment and health services received. Patient complaints can concern breaches to patient rights (e.g. waiting time, information and consent to treatment) or malpractice or harm to the patient (Winblad and Ringard, 2009). Complaint procedures can serve as one means of influencing the quality of healthcare through feedback to professional staff and organisations (Anell et al., 2012). These administrative bodies also gather information about and respond to the feelings and reactions of the patient and their next of kin (Winblad and Ringard, 2009).

According to the Patient Rights Act, patients in Norway can file a complaint to the county medical officer if they believe that they have not received the services they were entitled to. Complaints must be addressed to the person or organisation that made the decision, with a right of appeal to the county or the National Board of Health Supervision (Ringard et al., 2013).

The Danish National Agency for Patients' Rights and Complaints is responsible for dealing with complaints about professional treatment, the disregard of patient rights and decisions on compensation made by the Patient Insurance Association (Olejaz et al., 2012).

In Sweden, every county council has an advisory committee where people can address questions or complaints relating to medical treatment. The advisory committee is responsible for providing advice and for sorting out problems, reporting to professionals and services on any irregularities of significance to the patient (Anell et al., 2012). In cases of gross neglect or if an individual is anxious about reporting their complaints, an alternative is to apply to the government agency responsible for disciplinary measures (Anell et al., 2012).

In Finland, the Patient's charter states that patients who are dissatisfied with their care can submit a complaint to the director of the relevant provider organisation or alternatively to the regional or national authority (Finish National Supervisory Authority for Welfare and Health, 2012). In the case of patient injury during treatment, the Patient Insurance Centre can be consulted for compensation.

CURRENT STATE OF PLAY AND REMAINING CHALLENGES

As we have seen, quality and safety in health, welfare and social services is a political priority in the Nordic countries. There are unambiguous health policies emphasising patients' and their families' rights to be engaged and involved in their own treatment and care and in the development of health services. There are also a large number of patient organisations exercising influence on health policymaking and providing advice and support to patients and the public.

Health legislation complements and reinforces policies ensuring equal access to health and social services. Norway, Denmark and Finland have established separate bills, while Sweden uses multiple pieces of legislation to protect patients' rights. Choice of primary care provider (GP and healthcare centre/unit/hospital) has become mandatory in the four Nordic countries and there are waiting-time guarantees. Patients and citizens can easily access information about their rights and about health, disease and self-care and the quality of care on national websites. There are patient offices or patient ombudsmen in every region providing information and advice, and they play an important role in situations of complaints and redress.

There are few formal or legislative barriers preventing patients exercising choice in the Nordic countries (Vrangbæk et al., 2007). Nevertheless, a comparative study of Denmark, Norway and Sweden showed that patients exercised their right to choose hospitals to only a limited degree. Limitations in patients' knowledge about their right to choose, coupled with lack of information and support from their GP, were suggested as possible explanations (Vrangbæk et al., 2007). Ringard (2010) analysed Norwegian GPs' referral decisions for elective treatment. He found that GPs were most likely to send patients to a hospital other than their local one if they were in need of elective knee operations. Longer waiting times and distance from the local hospital increased the likelihood of an out-of-area referral. According to Glenngård et al. (2011), about 60% of the population in three Swedish counties felt that they had made a choice of provider following the introduction of the choice and privatisation legislation in 2010. The likelihood of choosing a provider increased with the introduction of new providers in the county and the perception of having enough information to make an informed choice. Those who preferred direct access to a specialist were less likely to choose to travel further for their care. However, people tended to choose providers they had previously been in contact with and many did not make active efforts to look for information.

Providing patients with easy access to information about their rights and the quality of care has been a key priority in the Nordic countries. In Sweden, the population survey (Befolkningsundersökning, 2014) reported a 10% increase in the proportion of respondents who said that they knew that they could call 1177 for advice and assistance. However, the extent to which people make use of this type of information is currently unknown (Anell et al., 2012). Glenngård et al. (2011) report that the two most common sources of information for people in the three Swedish counties responding to the survey were health providers followed by the Internet. People living in urban areas, those with higher education and younger people, were more likely to look for information on the Internet than other groups. The authors of this study noted the limitations of this type of search, making informed decisions about healthcare providers quite difficult. Information about symptoms and the benefits and harms of treatments is an essential prerequisite for shared decision-making, but there is a lack of web-based decision support to assist patients in choosing among treatment options (Docteur and Coulter, 2012).

National surveys of patient experiences with their hospital stay are conducted regularly in Norway, Sweden and Denmark. These indicate high levels of satisfaction with certain aspects of care, such as communications with health professionals, while also pointing to several priorities for improvement, for example involvement in decision-making, information and preparation for hospital discharge, co-ordination of care and accessibility of services (Bjerkan et al., 2014; Patientoplevelser Denmark, 2014; Swedish Patient Survey, 2014).

However, despite the progressive policies, it seems that in certain respects the Nordic countries may be lagging behind other western developed countries when it comes to measures of patients' experience. The Commonwealth Fund's 2011 International Health Policy survey, which include data from Norway and Sweden but not Denmark or Finland, pointed to information gaps during hospital discharge (Schoen et al., 2011), with the Nordic countries performing worse in this respect than most other countries in this eleven-country survey. In Sweden 67% of respondents and in Norway 71% reported the following problems:

1. Not given sufficient information about how to deal with their symptoms and when to seek further care
2. Not given information about who to contact for questions about their condition or treatment
3. Not given a written plan for care after discharge
4. No arrangements made for follow-up visits
5. Did not receive clear instructions about prescribed medication

In the United Kingdom, the comparative percentage was 26%. This evidence of poor performance is of concern because lack of preparedness and support for self-care can increase the risk of adverse events and avoidable readmissions to hospital.

Similarly, the 2014 International Health Policy Survey, which focused on older adults, suggested that the Nordic countries performed worse than many other countries in respect of access, co-ordination and patient-centred care (Osborn et al., 2014). There were less positive responses from people living in Norway and Sweden in response to questions about doctor–patient relationships, self-care of chronic conditions and end-of-life care:

- About 40% reported that their doctor always or often encouraged them to ask questions, while more than 80% reported this in the United States, Australia, France and Germany.
- Only 23% in Sweden and 27% in Norway reported that a healthcare professional discussed their main goals and gave instructions on symptoms to watch for in the past year, while almost 60% in the United Kingdom and the United States said they had had such discussions.
- Only 30% of respondents in Sweden and 20% in Norway said they had discussed their care preferences if they became unable to make decisions for themselves, while more than 70% of those in the United States and Germany had had such discussions.

Access to information about their condition, involvement in decisions about their care, support for self-management and encouragement of more conversations about end-of-life care are considered very important for enabling people to cope with chronic illness and to navigate the healthcare system effectively (Docteur and Coulter, 2012).

It is therefore of some concern that patients in these Nordic countries do not appear to be receiving sufficient support to empower them to make appropriate health choices.

CONCLUSION

Patient-centredness is an overarching aim for improving the quality of healthcare in the Nordic countries. Key attributes of a patient-centred health system are shared knowledge between professionals and patients, information to and involvement of patients in decision-making about treatment and care, transparency and access to information on system performance. Co-ordination and continuity of care are important, as are opportunities for collective engagement and co-design of service developments. The Nordic countries have introduced many progressive health policies, but there is still more to do to ensure that the systems are truly patient-centred. An important future direction is to strengthen the role of patients as active participants in their own care. Particular priorities include encouraging shared decision-making about treatments underpinned by appropriate information, together with personalised care planning and self-care support. Building health literacy among the population and encouraging health professionals to recognise the key role of patients and their families as co-producers of health will also be important in the next stage of health system reform.

REFERENCES

Andreassen, T.A., 2005. *Brukermedvirkning i helsetjenesten: Arbeid med brukerutvalg og andre medvirkningsprosesser.* [User involvement in healthcare service: Work with user committees and other participatory processes.] Oslo, Norway: Gyldendal akademisk.

Anell, A., Glenngård, A.H. and Merkur, S., 2012. Sweden: Health system review. *Health Systems in Transition*, 14(5):1–159.

Bate, P. and Robert, G., 2006. Experience based design: From redesigning the system around the patient to co-designing services with the patient. *Quality and Safety in Health Care*, 15(5):307–310.

Befolkningsundersökning. 2014. Vårdbarometern befolkningens attityder till, kunskaper om och förväntningar på hälso-och sjukvården. Sveriges kommuner och Landsting. http://www.vardbarometern.nu/ (accessed 21 October 2015).

Beresford, P., 2005. Developing the theoretical basis for service user/survivor-led research and equal involvement in research. *Epidemiologia e Psichiatria Sociale*, 14(1):4–9.

Berwick, D.M., 2009. What 'Patient-Centered' should mean: Confessions of an extremist. *Health Affairs*, 28(4):w555–w565.

Bjerkan, A.M., Holmboe, O. and Skudal, K.E., 2014. Pasienterfaringer med norske sjukehus 2013. Nasjonale resultater. [Patient experiences with Norwegian hospitals 2013.] PassOpp-rapport No. 2. Oslo, Norway: Nasjonalt kunnskapssenter for helsetjenesten.

Busch, T., Johnsen, E. and Vanebo, J.O., 2003. *Endringsledelse i det offentlige.* [Change management in public sector]. Oslo, Norway: Universitetsforlaget.

Cancer Society of Finland, 2016. Member organizations. [online] Available at https://www.cancersociety.fi/organisation/member-organisations/ (accessed 13 April, 2016).

Christensen, T. and Lægreid, P., 2001. New public management. The effects of contractualism and devolution on political control. *Public Management Review*, 3(1):73–94.

Coulter, A., 2011. *Engaging Patients in Healthcare.* Berkshire, England: Open University Press.

Coulter, A., Locock, L., Ziebland, S. and Calabrese, J., 2014. Collecting data on patient experience is not enough: They must be used to improve care. *British Medical Journal*, 348:g2225. doi:10.1136/bmj.g2225.

Crawford, M.J., Aldridge, T., Bhui, K., Rutter, D., Manley, C., Weaver, T., Tyrer, P. and Fulop, N., 2003. User involvement in the planning and delivery of mental health services: A cross-sectional survey of service users and providers. *Acta Psychiatrica Scandinavica*, 107(6):410–414.

Danish Health Act (sundhedsloven), 14 November, 2014. Danish Ministry of Health. Available through rets information https://www.retsinformation.dk/forms/r0710.aspx?id=152710 (accessed 10 April, 2016).

Danish Ministry of Health, 2015. National Quality Programme for Health 2015–2018. [pdf] Available through Danish Ministry of Health website http://www.sum.dk/Aktuelt/Publikationer/~/media/Filer%20-%20Publikationer_i_pdf/2015/Nationalt-kvalitetsprogram-for-sundhedsomraadet/Nationalt%20kvalitetsprogram%20for%20sundhedsområdet%20-%20april%202015.ashx (accessed 8 October 2015).

Danish Patients, 2016. About Danish Patients [online] Available at http://www.danskepatienter.dk/about-danish-patients (accessed 11 April, 2016).

Deilkås, E.C., Ingebrigtsen, T. and Ringård, Å., 2015. Norway, Chapter 24 in *Healthcare Reform, Quality and Safety. Perspectives, Participants, Partnerships and Prospects in 30 Countries*, Braithwaite, J., Matsuyama, Y., Mannion, R. and Johnson, J. (eds.). Farnham Surrey, England: Ashgate Publishing.

Docteur, E. and Coulter, A., 2012. Patient-centeredness in Sweden's health system. An external assessment and six steps process. Report 3, pp. 27–169. Available from Vårdanalys website http://www.vardanalys.se/Rapporter/2012/Patient-centeredness-in-Swedens-health-system---an-external-assesment-and-six-steps-for-progress/ (accessed 8 October 2015).

Doyle, C.L., Lennox, L. and Bell, D., 2013. A systematic review of evidence on the links between patient experience and clinical safety and effectiveness. *BMJ Open*, 3(1):1–18. doi: 10.1136/bmjopen-2012-001570.

Duncan, E., Best, C. and Hagen, S., 2010. Shared decision-making interventions for people with mental health conditions (review). *Cochrane Database of Systematic Reviews* (Online) No. 1.

Eriksen, E.O. and Weigård, J., 1993. Fra statsborger til kunde: kan relasjonen mellom innbyggerne og det offentlige reformuleres på grunnlag av nye roller? [From citizen to consumer: Can the relationship between citizens and public services be founded on new roles?] *Norsk Statsvitenskapelig Tidsskrift*, 9(2):111–131.

Finish Act on the Status and Rights of Patients No. 785/1992. Ministry of Social Affairs and Health, Finland. Available through Finlex Data Bank website http://www.finlex.fi/en/laki/kaannokset/1992/19920785 (accessed 8 October 2015).

Finish National Supervisory Authority for Welfare and Health, 2012. Patient's charter. Available through Valvira National Supervisory Authority for Welfare and Health website http://www.valvira.fi. (accessed 7 October 2015).

Glenngård, A.H., Anell, A. and Beckman, A., 2011. Choice of primary care provider: Results from a population survey in three Swedish counties. *Health Policy*, 103:31–37. doi:10.1016/j.healthpol.2011.05.014.

Haugum, M., Bjertnæs, Ø.A. and Lindahl, A.K., 2013. Pasientsikkerhet og kvalitet i helsetjenesten i 2013: en undersøkelse med basis i GallupPanelet. [Patient safety and quality in health services in 2013: A survey base don GallupPanelet.] Notat. Oslo, Norway: Kunnskapssenteret.

Helsenorge, 2016. Information in English. [online] Available at https://helsenorge.no/other-languages/english (accessed 13 April, 2016).

Institute of Medicine, 2001. *Crossing the Quality Chasm: A New Health System for the 21st Century*. Washington, DC: National Academy Press.

Infopankki.fi, 2016. Living in Finland. [online]. Available at http://www.infopankki.fi/en/living-in-finland/health (accessed 13 April, 2016).

Isaac, T., Zaslvsky, A.M., Cleary, P.D. and Landon, B.E., 2010. The relationship between patient' perception of care and measures of hospital quality and safety. *Health Services Research*, 45(4):1024–1040.

Kjellevold, A., 2005. Hensynet til brukeren – Idealet om brukerorientering i helse- og sosialtjenesten [The service user's interest – The ideal of user orientation in health- and social services], in *Brukernes medvirkning! Kvalitet og legitimitet i velferdstjenesten*, Willumsen, E. (ed.). Oslo, Norway: Universitetsforlaget, pp. 49–74.

Knudsen, J.L., Engel, C. and Eriksen, J., 2015, Denmark, Chapter 20 in *Healthcare Reform, Quality and Safety. Perspectives, Participants, Partnerships and Prospects in 30 Countries*, Braithwaite, J., Matsuyama, Y., Mannion, R. and Johnson, J. (eds.). Farnham Surrey, England: Ashgate Publishing.

Légaré, F., Ratté, S., Stacy, D., Kryworuchko, J., Gravel K., Graham, I.D. and Turncotte, S., 2010. Interventions for improving the adoption of shared decision making by healthcare professionals (review). *Cochrane Database of Systematic Reviews* (Online) No. 5.

Lewin, S., Skea, Z., Entwistle, V.A., Zwarenstein, M. and Dick, J., 2009. Intervention for providers to promote a patient-centered approach in clinical consultations (review). *Cochrane Database of Systematic Reviews* (Online) No. 1.

Mead, N. and Bower, P., 2000. Patient-centredness: A conceptual framework and review of the empirical literature. *Social Science and Medicine*, 51(7):1087–1110.

Norden, 2013. Nordisk Samarbejde på social- og sundhedsomårdet. Strategi for social og sundhedsområdet fra 2013 og frem: Nordisk Ministerråd.

Norwegian Directorate of Health, 2005. Og bedre skal det bli - Nasjonal strategi for kvalitetsforbedring i sosial- og helsetjensten (2005-2015). Til deg som leder og utøver. [And it's going to get better! National strategy for Quality Improvement in health and Social Services (2005–2015). For leaders and providers]. Oslo, Norway: Sosial og helsedirektoratet. Available from http://www.helsedirektoratet.no/kvalitetsforbedring/english/ (accessed 10 April, 2016).

Norwegian Directorate of Health, 2006. *Skjønnsmessig fastsettelse av forventet ventetid for rapportering til Fritt sykehusvalg Norge*. Oslo, Norway: Sosial- og helsedirektoratet. http://www.frittsykehusvalg.no/upload/Veileder%20IS-1200%20Forventet%20 ventetid-%20revisjon%2002022007.pdf (accessed 21 October 2015).

Norwegian Health Enterpris Act, 2001. Lov om helseforetak (Helseforetaksloven) 2001. Oslo, Norway: Ministry of Health and Care Services. Available from Lovdata website https://lovdata.no/dokument/NL/lov/2001-06-15-93 (accessed 10 April, 2016).

Norwegian Federation of Organizations of Disabled People (FFO), 2016. About FFO. [online] http://www.ffo.no/Organisasjonen/About-FFO/ (accessed 11 April, 2016).

Norwegian Ministry of Health and Care Services, 1996. *Åpenhet og helhet. Om psykiske lidelser og tjenestetilbudene* [Openness and wholeness: Mental health problems and service provision] St meld 25 (1996–1997). Oslo, Norway: Sosial- og helsedepartementet.

Norwegian Patient Rights Act, 1999. Lov om pasient- og brukerrettigheter (pasient- og brukerrettighetsloven) 1999. Oslo, Norway: Ministry of Health and Care Services. Available from Lovdata website https://lovdata.no/dokument/NL/lov/1999-07-02-63?q=Pasient+og+brukerrettighetsloven (accessed 10 April, 2016).

Ocloo, J.E. and Fulop, N., 2012. Developing a 'critical' approach to patient and public involvement in patient safety in the NHS: Learning lessons from other parts of the public sector. *Health Expectations*, 15:424–432. doi: 10.1111/j.1369-7625.2011.00695.x.

Olejaz, M., Juul Nielsen, A., Rundkjøbing, A., Okkels, B., Krasnik, A. and Hernández-Quevedo, C., 2012. Denmark: Health system review. *Health Systems in Transitions*, 14(2):1–192.

Osborn, R., Moulds, D., Squires, D., Doty, M. and Anderson, C., 2014. International survey of older adults finds shortcomings in access, coordination, and patient-centered care. *Health Affairs*, 33(12):2247–2255.

Øvretveit, J., Sachs, M.A. and Lindh, M., 2015. Sweden, Chapter 26 in *Healthcare Reform, Quality and Safety. Perspectives, Participants, Partnerships and Prospects in 30 Countries*, Braithwaite, J., Matsuyama, Y., Mannion, R. and Johnson, J. (eds.). Farnham Surrey, England: Ashgate Publishing.

Patientoplevelser, 2014. Udarbejdet af Enhed for Evaluering og Brugerinddragelse på vegne af regionerne. Available through Den Landsdækkende Undersøgelse af Patientoplevelser (LUP) website http://patientoplevelser.dk/sites/patientoplevelser.dk/files/dokumenter/filer/LUP/2014/lup_2014_national_rapport.pdf (accessed 7 October 2015).

Ringard, Å., 2010. Why do general practitioners abandon the local hospital? An analysis of referral decisions related to elective treatment. *Scandinavian Journal of Public Health*, 38:597–604.

Ringard, A., Sagan, A., Saunes, I.S. and Lindahl, A.K., 2013. Norway: Health system review. *Health Systems in Transition*, 15(8):1–162.

Schoen, C., Osborn, R, Squires, D., Doty, M., Pierson, R. and Applebaum, S., 2011. New 2011 survey of patients with complex care needs in eleven countries finds that care is often poorly coordinated. *Health Affairs*, 30(12):2437–2448.

Smith, E., Donovan, S., Beresford, P., Manthorpe, J., Brearley, S., Sitzia, J. and Ross, F., 2009. Getting ready for user involvement in a systematic review. *Health Expectations: An International Journal of Public Participation in Health Care and Health Policy*, 12(2):197–208.

Stamsø, M.A., 2005. New Public Management – Reformer i offentlig sektor [New Public Management – Reforms in public sector], in *Velferdsstaten i endring: norsk sosialpolitikk ved starten av et nytt århundre*, Stamsø, M.A. (ed.), pp. 67–85. Oslo, Norway: Gyldendal Akademisk.

Storm, M. and Edwards, A., 2013. Models of user involvement in the mental health context: Intentions and implementation challenges. *Psychiatric Quarterly*, 84(3):313–327.

Storm, M., Rennesund, Å.B. and Jensen, M.F., 2009. *Brukermedvirkning i psykisk helsearbeid.* [Service user involvement in mental health.] Oslo, Norway: Gyldendal akademisk.

Swedish Patient Survey, 2014. Nationell Patientenkät. Specialicerad sjukusvård 2014. Slutenvård. Sveriges kommuner och Landsting. Available from website http://npe.skl.se/ (accessed 10 April, 2016).

Swedish Rheumatism Association, 2016. About the Swedish Rheumatism association. [online]. Available at https://reumatikerforbundet.org/om-oss/about-swedish-rheumatism-association/ (accessed 13 April, 2016).

Toiviainen, H.K., Vuorenkoski, L.H. and Hemminki, E.K., 2010. Patient organizations in Finland: Increasing numbers and great variation. *Health Expectations*, 13:221–233.

Tritter, J.Q., 2009. Revolution or evolution: The challengers of conceptualizing patient and public involvement in a consumerist world. *Health Expectations*, 12:275–287.

Väntetider i vården, 2016. *Tillgänglig vård – en webbplats från Sveriges kommuner och landsting.* [online]. Available at http://www.vantetider.se/sv/ (accessed 13 April, 2016).

Vrangbæk, K., Østergren, K., Birk, H.O. and Winblad, U., 2007. Patient reactions to hospital choice in Norway, Denmark and Sweden. *Health Economics, Policy and Law*, 2(2):125–152.

Vuorenkoski, L., Mladovsky, P. and Mossialos, E., 2008. Finland: Health system review. *Health Systems in Transition*, 10(4):1–168.

1177 Vårdguiden, 2016. *Vårdguiden.* [online] Available at http://www.1177.se/Other-languages/Engelska/ (accessed 13 April, 2016).

Winblad, U. and Ringard, Å., 2009. Meeting rising public expectations: The changing role of patients and citizens, Chapter 6 in *Nordic Health Care Systems – Recent Reforms and Current Political Challenges*, Magnussen, J., Vrangbækk, K. and Saltman, R.B. (eds.). Maidenhead, England: Open University Press.

3 Studying Patient Safety and Quality from Different Methodological Angles and Perspectives

Siri Wiig and Tanja Manser

CONTENTS

INTRODUCTION

Patient safety and quality in healthcare has become an international priority involving research programmes and policy reforms to build safer and better healthcare systems (e.g. WHO, 2008; Iedema, 2009; Jha et al., 2010; Shekelle et al., 2011; Mitchell et al., 2015). This also applies to the Nordic context, with the Nordic countries having both separate country-specific strategies and cross-national strategic collaborations to improve patient safety and quality in healthcare (Norden, 2013). Several initiatives aiming at structural (e.g. regulation, accreditation, national error reporting system) and cultural improvement (e.g. patient safety culture measurement, leadership walk rounds) have been undertaken already and will further be implemented in the years to come (e.g. Vuorenkoski, 2008; Anell et al., 2012; Meld.St nr. 10, 2012–2013; Olejaz et al., 2012; Saunes and Ringaard, 2013; Sigurgeirsdóttir et al., 2014).

Understanding patient safety and quality improvement interventions and mechanisms within complex healthcare systems has become an increasingly important research field (Øvretveit, 2014; Mason et al., 2015; Portela et al., 2015). Research is a key component of improving patient safety and quality (Pronovost et al., 2015) as it creates an in-depth understanding of the problems and thus a basis for effective and sustainable improvement. The World Health Organization (WHO) has suggested that the top five research priorities among developed countries, such as the Nordic countries, are related to

1. Lack of communication and co-ordination (including discontinuity and co-ordination across organisations)
2. Latent errors and organisational failures
3. Poor safety culture and blame-oriented processes
4. Cost-effectiveness of risk reduction strategies
5. Developing better safety indicators (WHO, 2008; Bates et al., 2009)

Despite prolonged effort, the field is improving slowly, which is partly due to limited evidence for the development and dissemination of successful practices (Pronovost et al., 2009; Shekelle et al., 2011). In 2011, an Agency of Healthcare Research and Quality expert group argued that research approaches require more extensive use of theory; a more detailed description of interventions and their implementation; enhanced explanation of desired and undesired outcomes and improved description and measurement of context and how context influences improvement interventions (Shekelle et al., 2011).

While research is crucial to improving patient safety and quality, there is increasing concern that research strategies and methods are not sufficiently adapted to the complexities of healthcare. Some claim that too much attention has been paid to the formal and structural dimensions at the expense of the social and affective dimensions relevant to patient safety and quality (Bate et al., 2008; Idema, 2009; Wiig et al., 2014a). Since the role of social processes is under-researched, there is a need for approaches that enable to grasp the interactive processes of establishing and sustaining patient safety and quality.

AIM AND LIMITATIONS

This chapter reflects on the challenges of researching patient safety and quality in healthcare (Brown et al., 2008; Pronovost et al., 2009; Portela et al., 2015) and on methodological choices such as the selection of appropriate theoretical frameworks, research design, data collection methods, context mapping, analytical tools, understanding interventions and explaining results. The aim is to discuss the characteristics of the healthcare research context and the implications of different methodological perspectives.

In doing so, we assume that readers have a good understanding of most research methods and designs (e.g. Denzin and Lincoln, 1994; Reason and Bradbury, 2001; Yin, 2004; Creswell, 2014). Hence, we will only briefly cover the landscape of methods and perspectives.

LANDSCAPE OF METHODS AND PERSPECTIVES

The landscape of research methods and perspectives that have been applied to patient safety and quality in healthcare is large and sprawling. It varies depending on multiple factors such as philosophy of science positioning, professional and academic background, theoretical perspectives and level of interest (Fulop et al., 2001). For example, understanding of patient safety risk may vary from a positivist view conceptualising risk as a physically given attribute where objective facts can be explained, predicted and controlled to a constructivist view arguing that risk is a socially constructed phenomenon (e.g. Renn, 1992, 2007).

Methodological approaches may vary from quantitative designs relying on the 'gold standard' of randomised controlled trial designs focusing on direct inferences on causality on the one hand to qualitative designs and ethnographic approaches on the other hand focusing on in-depth inquiry aiming to understand work practices, culture, processes, social interaction, meanings of the social world and why and how changes appear (Fulop et al., 2001; Portela et al., 2015). Moreover, a wide range of theoretical perspectives and models (Davidoff et al., 2015) can be applied to guide research from, for example, simple linear to complex safety models developed within the safety sciences (Hollnagel, 2004; Hollnagel et al., 2013). Within the tradition of quality improvement, different models and tools (e.g. Total Quality Management, Lean thinking, Six Sigma) have been widely used to guide quality improvement in healthcare organisations (Øvretveit and Staines, 2007; Bate et al., 2008; Powell et al., 2009; Robert et al., 2011; Wiig et al., 2014a). These theories and models are not always well defined and healthcare organisations often draw on a range of tools and principles from different approaches (Powell et al., 2009). Although a wide range of theories exist within patient safety and quality (see also Chapter 4), the role and value of theory is still under-recognised and mystified in healthcare. Davidoff et al. (2015) propose that more informed use of theory could strengthen improvement programmes and facilitate the evaluation of their effectiveness.

Patient safety and quality in healthcare is not the province of any one discipline, research design or method of measurement, therefore understanding and improvement require a collaborative synthesis of different perspectives and contributions (Walshe and Boden, 2006; Pronovost et al., 2015). Research aiming at understanding the causes of adverse events and patient harm is complex, because the events often relate to multiple factors (e.g. communication, teamwork, care process design, instrumentation and facilities, human resources, service organisation, management and leadership). According to Walshe and Boden (2006), a range of disciplines, including psychology, sociology, clinical epidemiology, quality management, technology and informatics and the law, all have important parts to play in approaching and analysing errors and adverse events. The advantage of different disciplines and backgrounds is that they foster diversity and complementary perspectives and conceptual understandings of patient safety and quality.

The choice of level of inquiry constitutes an important aspect within the methodological landscape. Different schools of thought focus their study on the same research topic but at different organisational or system level. The choice of level is partly informed by theory and partly by pragmatism (Fulop et al., 2001). Level of interest may vary from studies of a single level involving either the clinical micro-level (clinical team), the meso-level (organisational level) or the macro-level (national level) to multi-level approaches including several or all levels and the role of interaction and interfaces between them. Also, depending on the chosen level of inquiry, the assignment of these levels is fluid ranging from the individual clinician to cross-country comparisons. As a general rule of thumb, multi-level approaches should always include one level above and below the focal level of inquiry. However, the current research usually includes a single care sector (e.g. primary or specialised healthcare services) and seldom embraces a multi-level perspective (Wiig, 2008; Robert et al., 2011).

THREE CONTEXTUAL ISSUES CHALLENGING RESEARCH ON PATIENT SAFETY AND QUALITY

In the following, we highlight three specific characteristics of the healthcare context and how they challenge research. First, we focus on how patient safety and quality are conceptualised; second, we point out the changing nature of the healthcare system and third, we highlight the role of patients and patient involvement in healthcare.

Conceptualising Patient Safety and Quality

Research methods need to capture how patient safety and quality are understood, created, enacted and sustained. Conceptualising patient safety and quality thus constitutes a challenge that research needs to approach.

Looking at patient safety, Vincent (2006: 14) argues that it is concerned primarily with the avoidance, prevention and amelioration of adverse outcomes or injuries stemming from healthcare itself. Safety does not reside in a person, device or department, but emerges from the processes and the interfaces in the socio-technical system of healthcare. This implies that safety depends on interaction between multiple actors, technologies and processes (e.g. Robert et al., 2011; Wiig et al., 2014a; Waterson et al., 2015). Having this in mind, Pronovost et al. (2015) argue that healthcare researchers so far have been unable to capture the complexity of patient safety conceptualisation, because of lacking transdisciplinary approaches (i.e. multidisciplinary combined applied and basic research principles). They argue that many healthcare researchers tend to consciously and narrowly focus on conceptualisations in isolation rather than considering them as a result of several interdependent conceptualisations (systems at work). Too often, research focuses on mono-disciplinary approaches adopting a linear approach from basic to clinical research. What is missing is a systematic, sector-wide approach, underpinned by conceptualisations of safety science adopted in other industries but not yet in healthcare (Pronovost et al., 2015).

Looking at quality, conceptualisation is also complex. Diverse definitions of quality have been developed and often include the six dimensions of effectiveness, timeliness, safety, equity, efficiency and patient centredness. Here, patient safety is conceptualised as a sub-dimension of quality. Some of these dimensions may be inherently conflicting in practical settings (e.g. improved safety might reduce patient centredness or efficiency) (Wiig et al., 2013), and the different quality dimensions may require different methodological approaches to collect and analyse data. Hence, including all dimensions in research activities is a challenging methodological task. It implies that researchers sometimes limit the number of included quality dimensions, in order to be able to go into details. A study of quality conceptualisations across European hospitals demonstrated how a delimitation of the quality conceptualisation in the research constituted a challenge for the data collection and analysis (Wiig et al., 2014b). The 'narrow' conceptualisation of quality used (clinical effectiveness, patient safety and patient experience) evolved into a limitation in the analysis demonstrating how other quality dimensions emerged from the data (efficiency, time, economic issues). A broader conceptualisation of quality would have enabled a more thorough and consistent exploration of the data. However, this study demonstrated that the quality conceptualisations differed across system levels (macro–meso–micro), among professional groups (nurses, doctors, managers) and between the micro systems (Wiig et al., 2014b).

Further, patient safety and quality can be conceptualised as moving targets (Vincent and Amalberti, 2015) since meanings and concepts are dynamic and dependent upon innovation and improvement within the healthcare context itself. What was accepted as sound professional practice 10 years ago might not be considered acceptable by professional or regulatory bodies today. As standards improve and concerns for patient safety and quality increase, an increasing number of adverse events are, for example, regarded as preventable. Vincent and Amalberti (2015) argue that innovations and improving standards alter our conceptualisation of both harm and preventability in healthcare. From a practical, conceptual and methodological perspective, this is challenging as the definition of what constitutes patient harm or safe care is becoming difficult to pin down since a growing amount of events with changing characteristics are labelled a safety issue. The same argument is also relevant for quality standards and acceptable level of care quality (Braut, 2011).

A Continuously Changing Context

Healthcare is characterised by the need to continuously address changes either imposed on the system, team or individual healthcare professionals by governmental bodies (e.g. national reforms and standards), technological or professional development (e.g. new knowledge, procedures, competence or diagnostic tools) or driven from an internal need for organisational change (e.g. restructuring or downsizing) or driven by patients (e.g. demands for involvement) to mention some. Understanding how patient safety and quality processes are affected by these change processes is a challenging research task, and the same applies for evaluating interventions implemented within these systems. Interventions occur in contexts that are difficult to describe in detail in terms of how and what factors influence the results. This makes it difficult for others to understand and repeat the intervention (Øvretveit, 2014). A prominent example is the introduction of the safe surgery checklist that may fail to achieve the intended improvements if not accompanied by other elements such as staff training and leadership support. Often, we learn from a failed replication attempt and the differences between contexts.

Another feature of a continuously changing context is the care process itself. The increasing degree of specialisation in healthcare combined with the demands for treatment at the lowest possible care level causes a growing number of handovers. These changes imply that patients are transferred between and within service levels (primary and acute services), services types (intensive care, cancer care) and teams. Previous research states that – across different healthcare settings – current handover processes are highly variable and potentially unreliable (Manser and Foster, 2011). More knowledge is required to understand these contextual changes and how they affect patient safety and quality. This requires methodological approaches and perspectives that manage to map change and care processes along different patient pathways and trajectories.

Patient Involvement

Even though patient and public involvement (PPI) has become part of everyday rhetoric in many countries, it often proofs difficult in practice (Coulter, 2011). A Norwegian study of patient experiences in quality improvement in hospitals demonstrated that macro-level policymakers had wide-ranging expectations for the integration of patient experiences. Results nevertheless demonstrated a lack of expertise in Norwegian hospitals of adapting and implementing tools and methods for using patient experiences as a resource in quality improvement (Wiig et al., 2013). Patients and family can play a distinct role in improving patient safety and quality by, for example, choosing safe healthcare providers, observe and check care processes, identify and report complications and adverse events, give feedback and advocate attention to patient safety and quality issues (Coulter, 2011). Sutton et al. (2015) found three main types on patient involvement in patient safety in their literature review relating to (1) monitoring treatment and speaking up, (2) patient perception of safety and experiences of raising concerns and (3) patient feedback and error reporting. Another systematic review of patient involvement in patient safety revealed three different themes relating to (1) satisfaction with and need for knowledge about healthcare and the health system, (2) sharing responsibility and accountability for safety and (3) the need to overcome language barrier to prevent harm and error (Severinson and Holm, 2015). These two reviews demonstrate different themes and activities related to patient involvement in patient safety at both individual and collective level that research needs to explore in more detail.

In terms of patient involvement in quality improvement, efforts often focus on technicalities of involvement such as training of patients and healthcare professionals to strengthen their skills and technical knowledge (e.g. be more articulate, able to 'represent') (Renedo et al., 2015). Other PPI methods to involve patients, beyond educational programmes, are, for example, family meetings, discharge plans in transitional care, checklists or home visits (Dyrstad et al., 2014). However, the role of organisational context in supporting PPI is rarely analysed, although elements of organisational culture could be important to mediate success. In an ethnographic study investigating PPI activities in England, Renedo et al. (2015)

found four key elements in the organisational culture that contributed to successful PPI in quality improvement:

1. Emphasis on non-hierarchical, multidisciplinary collaboration between patients and professionals
2. Staff ability to model desired improvement and PPI behaviours of mutual recognition and respect
3. Commitment to rapid improvement to ensure translation of research to practice
4. Constant and iterative process of collecting data and reflection facilitated by the use of quality improvement methods and the commitment to act on that learning (Renedo et al., 2015)

The growing influential role of patients and their families in healthcare adds complexity to patient safety and quality research. It implies a need for methodological angles and perspectives that acknowledge the role of the patient and the family (Coulter, 2011) by also engaging them actively in the research process. The previous research highlights 'soft' dimensions related to patient involvement, such as respect, commitment, accountability and responsibility that researchers and practitioners need to address more thoroughly than today.

DISCUSSION

As outlined earlier, the healthcare context represents several research challenges. In this section, we will discuss the role of different methodological perspectives in overcoming these challenges.

Research Addressing Conceptualisation of Patient Safety and Quality

Since patient safety and quality concepts and dimensions differ and are interrelated (Doyle et al., 2013), they require different angles and methods to collect data. For example, studying the two dimensions, cost-effectiveness of patient safety improvement, researchers need to collect data from economic figures, calculate intervention costs and map decrease or increase in the defined variables. According to Recende et al.'s (2012) review of methods used in economic evaluation in patient safety, costs are rarely evaluated as part of implementing patient safety improvement measures. The review demonstrates a narrow perspective on the costs related to a few adverse event types (healthcare-associated infections/infection control activities and medication-related adverse events). Recente et al. (2012) recommend stronger emphasis on transparency, data collection and methodological choices. Research programmes in the EU and the Nordic counties often require economic evaluations of interventions, but as demonstrated in the review this is complicated. We believe that a broader perspective in economic evaluations should be applied. The evaluations could be supplemented by using methods such as focus groups, semi-structured interviews to explore, for example, senior managers' perspective on resources invested in improvement efforts over time such as accreditation programmes, change in staffing level or establishment of learning arenas. Also, questionnaires could support the understanding of economic consequences of improvements measures. Analysed together with commonly used economic evaluations in a mixed methods approach they can contribute to a broader conceptualisation of resources invested in patient safety and quality, beyond the single perspective of adverse events. This discussion also relates to what is conceptualised as evidence in research on patient safety and quality. Is it merely outcome measures (e.g. infection rate, mortality) or could process measures provide valuable information on the relation between patient safety, quality and costs? Since research on patient safety and quality is still in its early stages, it needs both.

Despite the complexity of the patient safety and quality conceptualisations in healthcare improvement, healthcare generally operates in silos. Improvement efforts are often superficial, fragmented and not well co-ordinated. To overcome this issue, there has been a call for more multidisciplinary research approaches (Pronovost et al., 2015). Different disciplines contribute with different conceptualisations fostering a broader perspective in research. Within organisational sciences, the concept of requisite variety is used when discussing the need for a divergence in analytical perspectives among members required to understand the complexity of a system (e.g. Weick, 1987). Applied to research, this means research teams need enough analytical variety to conceptualise important information, diagnose the problem and suggest relevant improvement measures. But there are also potential obstacles to multidisciplinary research, which are often not discussed. Disciplines and traditions within disciplines stem from different paradigms and theories of knowledge. Clearly, these differences are important and are likely to influence a broad range of aspects relevant to the research, such as types of research undertaken, focus on process or outcome, the role of context and generalisability and the types of methods. It is suggested that multidisciplinary research is easiest when researchers from different disciplines share paradigms or represent a less extreme end of perspectives working together (Fulop et al., 2001). However, while a common language and general openness towards other perspectives is critical, analytical conceptualisation still needs to be variable enough to actually benefit from diversity.

Research Addressing Changing Contexts

Continuous change challenges some research methods and favours others. The choice of methods depends on the aim and research problem (explorative, descriptive, explanatory or experimental). For those methods exploring a phenomenon in its context, such as case studies or action research, the change processes constitute part of the context that the research tries to understand. Instead of controlling influencing factors or confounders, researchers then focus on providing an in-depth description of the research context, helping others to assess if findings are relevant in another context. Such studies provide valuable insights into context characteristics that can be included in explanatory studies.

Patient safety and quality research also needs to explore the consequences of changes implemented at different levels, such as reforms or organisational restructuring. Improvement efforts in healthcare systems undergoing fundamental restructuring require greater understanding of how organisational context affects outcomes (e.g. Aiken et al., 1997) and processes. At the national level, several Nordic countries implement system-wide reforms (e.g. the co-ordination reform described in Chapter 10 or national clinical registries in Chapter 7) that require scientific evaluation of implications (structure, process, outcome), the implementation process and strengths and weaknesses (Øvretveit, 2014). Such research evidence should inform stakeholders' future decision-making.

The changes into more highly specialised and centralised hospital services in the Nordic countries will require more cross-country collaboration (Norden, 2013), with the potential of adding handover and fragmentation to service provision. Manser (2013) discusses this in a critical reflection of current research approaches to patient handover and criticises the current predominant research strategy to study isolated handover episodes in studying this changing context. This strategy replicates the problem of fragmentation of care that we aim to overcome by improving handover. Thus, there is a need for a patient-centred approach to handover research investigating the interdependencies of handovers occurring along the care path. The patient and the family are the only ones with a complete experience and picture of the patient journey and research methods need to integrate these experiences in the research to understand the implications of the changing context on patient safety and quality. Possible approaches may be both qualitatively and quantitatively oriented by using, for example, observations of patients through the healthcare system over time, interviews with patients and family along the pathway, surveys with several measurements to map involvement at different stages of the care process or a combination of methods. Such approaches may contribute to novel insights and help increase the effectiveness and sustainability of interventions to improve handover (Manser, 2013: 194) in both a Nordic and an international perspective.

Research Addressing Patient Involvement

Most healthcare research would be impossible without active involvement of patients. Patient involvement in research ranges from involvement in only one part of the study to conducting and controlling the entire study. There is limited evidence that patient involvement improves the quality and impact of research (Coulter, 2011; Staley, 2015). However, the patient perspective and patient involvement should be considered integral to research on patient safety and quality to further develop our understanding of the systems and challenges in service provision. This means, for example, including patients as informants, stakeholders, members in advisory groups and co-researchers adding to the multidisciplinary perspectives held by the researchers and clinicians. Within patient safety research, it is also important to include the voice of the harmed patients (Ocloo, 2010). The latter is challenging, as patients might no longer be alive, and next of kin in grief might have to be approached by researchers. The demand for future studies on patient safety and quality in healthcare will require methods that enable grasping both patient involvement and the patient perspective to avoid a fragmented (e.g. only one stakeholder in the care pathway) or a one-sided picture (e.g. only the hospital/health professional perspective) of patient safety and quality. Angel and Frederiksens' (2015) recent review on patient involvement concludes that empirical studies do not consider whether it is at all possible to overcome the embedded societal structures in which patients must participate. Involvement in its ideal form cannot be achieved, they argue, due to the unequal relationship between patients and professionals. A suggested way forward is to consider patient involvement on a case-by-case basis, based on patients' situation and professionals' attitudes (Angel and Frederiksen, 2015: 1536).

To overcome the narrow focus on technicalities of PPI and explore the soft dimensions (e.g. respect, commitment, accountability and responsibility), the study by Renedo et al. (2015) offers an interesting angle. Their study approached the role of organisational context and culture for PPI by using a large ethnographic study of PPI activities. They conducted in-depth interviews with patients and carers involved in improvement projects and spent numerous hours observing PPI activities to investigate how PPI was organised and enacted in practice. Moreover, observation included meetings where teams of healthcare professionals, researchers and patient participants came together to work in improvement projects and events to facilitate learning about quality improvement methods and PPI. This illustrates research focusing on a high degree of qualitative methods including a collaborative component between researchers, patients and healthcare professionals to understand processes and practical aspects of patient involvement: what happens when people participate, what kind of actions do patients engage with as part of their involvement and what is the nature of professional–patient interactions. These topics cannot be explored by traditional RCT design and strict emphasis on causality. This is a developing research area and future research needs to focus on elements of, for example, trust, responsibility and power issues (Severinson and Holm, 2015), but as it becomes more mature there is need for methods to provide evidence of the effect of patient participation on patient safety and quality processes, structures and outcome.

WHAT IS NEXT IN THE NORDIC CONTEXT?

We need research perspectives and methods that are able to conceptualise the complex characteristics of patient safety and quality as continuously moving targets. The complexity of the context and the concepts call for research approaches integrating a variety of methods (qualitative, quantitative, mixed methods) that build on each other to explore and explain the different conceptual dimension and the emergence of patient safety and quality. This requires multidisciplinary research approaches, cultures in healthcare valuing diversity and research programmes at national and Nordic level requesting collaboration between different disciplines and stakeholders.

The Nordic model and values (described in Chapter 1) are challenged, because of three megatrends: the digital revolution, globalisation and increasing life expectancy (Norden, 2014). Hence, the Nordic healthcare context faces similar challenges of change involving a growing population of elderly, while the predicted number of healthcare professionals is too low. Significant changes are expected in the way healthcare is delivered and e-health and telecare are expected to have an increasingly important role (Norden, 2013; Guise et al., 2014; Wiig et al., 2014c). The Nordic research community should be closely coupled to the implementation of changes in at least two ways: first, by research approaches focusing on foresight and scenario building and consequence analysis to predict future patient safety and quality implications, avoid pitfalls and take action in light of possible and desirable future outcome (Durance and Godet, 2010; Ringland, 2010) and, second, by evaluating patient safety and quality implications of the change processes over time by using a variety of methods (qualitative, quantitative, mixed methods) and data sources (registry data, indicators, patient-reported outcome, patient and health professionals, healthcare managers and policymakers).

Last but not least, future research should include patients and family members not only as a valuable information and data source, but as partners in research. Researchers need to identify innovative approaches to integrate patients and families as partners to inform research questions and methods and develop the research team's capability of targeting the different conceptualisations of patient safety and quality. The Nordic countries and the Nordic model (Karlsen and Lindøe, 2006) appear as a possible context that could integrate patient involvement in research. Being characterised by a low level of hierarchy, tripartite collaboration, equal opportunities and a large public sector that provides its citizens with generous benefits, welfare services and social safety net (Karlsen and Lindøe, 2006; Norden, 2014), the Nordic context should foster patient involvement in patient safety and quality research.

REFERENCES

Aiken, L., Sochalski, J. and Lake, E.T. (1997). Studying outcomes of organizational change in health services. *Medical Care*, 35(11 Supplement), NS6–NS18.

Anell, A., Glenngård, A.H. and Merkur, S. (2012). Sweden health system review. *Health Systems in Transition*, 14(5), 1–159.

Angel, S. and Frederiksen, K.N. (2015). Challenges in achieving patient participation: A review of how patient participation is addressed in empirical studies. *International Journal of Nursing Studies*, 52, 1525–1538.

Bate, P., Mendel, P. and Robert, G. (2008). *Organizing for Quality*. Oxford, U.K.: Radcliffe Publishing.

Bates, D.W., Larizgoitia, I., Prasopa-Plaizier, N., Jha, A., on behalf of the Research Priority Setting Working Group of the World Alliance for Patient Safety. (2009). Global priorities for patient safety research. *British Medical Journal*, 338, b1775.

Braut, G.S. (2011). Legal requirements related to governance and health services. In Molven, O. and Ferkis, J. (eds.), *Healthcare, Welfare and Law*. Oslo, Norway: Gyldendal Akademisk.

Brown, C., Hofer, T., Johal, A., Thomson, R., Nicoll, J., Franklin, B.D. and Lilford, R.J. (2008). An epistemology of patient safety research: A framework for study design and interpretation. Part 2. Study design. *Quality and Safety in Healthcare*, 17, 163–169.

Creswell, J. (2014). *Research Design: Qualitative, Quantitative, and Mixed Methods Approaches*. Los Angeles, CA: Sage Publications.

Coulter, A. (2011). *Engaging Patients in Healthcare*. Berkshire, England: Open University Press.

Davidoff, F., Dixon-Woods, M., Leviton, L. and Michie, S. (2015). Demystifying theory and its use in improvement. *BMJ Quality and Safety*, 24, 228–238.

Denzin, N.K. and Lincoln, Y.S. (eds.) (1994). *Handbook of Qualitative Research*. Thousand Oaks, CA: Sage Publications.

Doyle, C.L., Lennox, L. and Bell, D. (2013). A systematic review of evidence on the links between patient experience and clinical safety and effectiveness. *BMJ Open*, 3(1), e001570.1–19.

Durance, P. and Godet, M. (2010). Scenario building: Uses and abuses. *Technological Forecasting & Social Change*, 77, 1488–1492.

Dyrstad, D.N., Testad, I., Aase, K. and Storm, M. (2014). A review of the literature on patient participation in transitions of the elderly. *Cognition, Technology & Work*, 17, 15–34.

Fulop, N., Allen, P., Clarke, A. and Black, N. (eds.) (2001). *Studying the Organization and Delivery of Health Services: Research Methods*. London, England: Routledge.

Guise, V., Anderson, J. and Wiig, S. (2014). Patient safety risks associated with telecare: A systematic review and narrative synthesis of the literature. *BMC Health Services Research*, 14, 588. doi: 10.1186/s12913-014-0588-z. http://www.biomedcentral.com/1472-6963/14/588.

Hollnagel, E. (2004). *Barriers and Accident Prevention*. Aldershot, U.K.: Ashgate.

Hollnagel, E., Braithewaite, J. and Wears, R. (2013). *Resilient Health Care*. Surrey, U.K.: Ashgate.

Iedema, R. (2009). New approaches to researching patient safety. *Social Science & Medicine*, 69, 1701–1704.

Jha, A.K., Prasopa-Plaizier, N., Larizgoitia, I., Bates, D.W., on behalf of the Research Priority Setting Working Group of the WHO World Alliance for Patient Safety. (2010). Patient safety research: An overview of global evidence. *Quality and Safety in Health Care*, 19, 42–47.

Karlsen, J.E. and Lindøe, P.H. (2006). The Nordic OSH model at a turning point? *Policy and Practice in Health and Safety*, 14(1), 17–30.

Manser, T. (2013). Fragmentation of patient safety research: A critical reflection of current human factors approaches to patient handover. *Journal of Public Health Research*, 2(e33), 194–197.

Manser, T. and Foster, S. (2011). Effective handover communication: An overview of research and improvement efforts. *Best Practice & Research Clinical Anaesthesiology*, 25(2), 181–191.

Mason, S.E., Nicolay, C.R. and Darzi, A. (2015). The use of Lean and Six Sigma methodologies in surgery: A systematic review. *The Surgeon*, 13, 91–100.

Mitchell, I., Schuster, A., Smith, K., Pronovost, P. and Wu, A. 2016. Patient safety reporting: A qualitative study of thoughts and perceptions of experts 15 years after '*To Err is Human*'. *BMJ Quality and Safety*, 25, 92–99.

Meld. St nr. 10. (2012–2013). *God kvalitet og trygge tjenester.* Helse og Omsorgsdepartementet. (Report to the Parliament: High quality and safe services).

Norden. (2013). *Nordisk Samarbejde på social- og sundhedsområdet. Strategi for social og sundhedsområdet fra 2013 og frem*. København, Denmark: Nordisk Ministerråd. ISBN:978-92-893-2578-3. (Nordic Collaboration on Social Affairs and Health. Strategy from 2013 and further on).

Norden. (2014). Does the Nordic Model need to change?: Overview of the research report, The Nordic model – Challenged but capable of reform. Copenhagen, Denmark: Nordic Council of Ministers. ISBN: 978-92-893-3884-4.

Ocloo, J.E. (2010). Harmed patients gaining voice: Challenging dominant perspectives in the construction of medical harm and patient safety reforms. *Social Science & Medicine*, 71(3), 510–516.

Olejaz, M., Juul Nielsen, A., Rudkjøbing, A., Okkels Birk, H., Krasnik, A. and Hernández-Quevedo, C. (2012). Denmark health system review. *Health Systems in Transition*, 14(2), i–xxii, 1–192.

Øvretveit, J. (2014). *Evaluation Improvement and Implementation for Health*. Berkshire, England: McGraw-Hill Education, Open University Press.

Øvretveit, J. and Staines, A. (2007). Sustained improvement? Findings from an independent case study of the Jönköping quality programme. *Quality Management in Healthcare*, 16, 68–83.

Portela, M.C., Pronovost, P., Woodcock, T., Carter, P. and Dixon-Woods, M. (2015). How to study improvement interventions: A brief overview of different study types. *BMJ Quality and Safety*, 24, 325–336.

Pronovost, P., Goeschel, C.A., Marsteller, J., Sexton, B., Pham, J. and Berenholtz, S. (2009). Framework for patient safety research and improvement. *Circulation*, 119, 330–337.

Pronovost, P., Ravitz, A.D., Stoll, R.A. and Kennedy, S.B. (2015). Transforming patient safety. A sector-wide systems approach. Report of the *WISH Patient Safety Forum 2015*. World Innovation Summit for Health.

Powell, A.E., Rushmer, R.K. and Davies, H.T.O. (2009). A systematic narrative review of quality improvement models in health care. Dundee, U.K.: Social Dimensions of Health Institute, The Universities of Dundee and St Andrews.

Reason, P. and Bradbury, H. (2001). *Handbook of Action Research: Participative Inquiry and Practice*. London, England: Sage Publications.

Recende, B., Or, Z., Com-Ruelle, L. and Michel, P. (2012). Economic evaluations in patient safety: A literature review of methods. *BMJ Quality and Safety*, 21, 457–465.

Renedo, A., Marston C.A., Spyridonidis, D. and Barlow, J. (2015). Patient and public involvement in healthcare quality improvement: How organizations can help patients and professionals to collaborate. *Public Management Review*, 17(1), 17–34.

Renn, O. (1992). The concept of risk. In Krimsky, S. and Golding, D. (eds.), *Social Theories of Risk*. Westport, CT: Praeger.

Renn, O. (2007). *Risk Governance*. London, England: Earthscan.

Ringland, G. (2010). The role of scenarios in strategic foresight. *Technological Forecasting & Social Change*, 77, 1493–1498.

Robert, G.B., Anderson, J.E., Burnett, S.J. et al. (2011). A longitudinal, multi-level comparative study of quality and safety in European hospitals: The QUASER study protocol. *BMC Health Services Research*, 11, 285. Online access: http://www.biomedcentral.com/1472-6963/11/285.

Saunes, I. and Ringard, Aa. (2013). What is done to improve patient safety? Initiatives in seven countries. Report from the Norwegian Knowledge Centre for the Healthcare Services, Oslo, Norway. Report nr 16-2013. ISBN 978-82-8121-637-2.

Severinson, E. and Holm, A.L. (2015). Patients' role in their own safety – A systematic review of patient involvement in safety. *Open Nursing Journal*, 5, 642–653.

Sigurgeirsdóttir, S., Waagfjörð, J. and Maresso, A. (2014). Iceland health system review. *Health Systems in Transition*, 16(6), 1–182, xv.

Shekelle, P.G., Pronovost, P.J., Wachter, R.M. et al (2011). Advancing the science of patient safety. *Annals of Internal Medicine*, 154(10), 639–696.

Staley, K. (2015). 'Is it worth doing?' Measuring the impact of patient and public involvement in research. *Research Involvement and Engagement*, 1, 6.

Sutton, E., Eborall, H. and Martin, G. (2015). Patient involvement in patient safety: Current experiences, insights from the wider literature, promising opportunities? *Public Management Review*, 17(1), 72–89.

Vincent, C. (2006). *Patient Safety*. Edinburgh, U.K.: Elsevier.

Vincent, C. and Amalberti, R. (2015). Safety in healthcare is a moving target. *BMJ Quality and Safety*, 24, 539–540.

Vuorenkoski, L. (2008). Finland health system review. *Health Systems in Transition*, 10(4): 1–170.

Walshe, K. and Boden, R. (eds.) (2006). Introduction. In *Patient Safety: Research into Practice*. Berkshire, U.K.: Open University Press.

Waterson, P., Robertson, M., Cooke, N. et al. (2015). Defining methodological challenges and opportunities for an effective science of socio technical systems and safety. *Ergonomics*, 58(4), 565–599.

Weick, K. (1987). Organizational culture as a source of high reliability. *California Management Review,* XXIX (2, Winter), 112–127.

WHO. (2008). World alliance for patient safety. Summary of the evidence on patient safety: Implications for research. Geneva, Switzerland: World Health Organization.

Wiig, S. (2008). Contribution to risk management in the public sector. No. 48. PhD thesis, University of Stavanger, Stavanger, Norway.

Wiig, S., Aase, K., von Plessen, C., Burnett, S., Nunes, F., Weggelaar, A.M., Anderson-Gare, B., Calltorp, J., Fulop, N., for QUASER Team. (2014b). Talking about quality: Exploring how 'quality' is conceptualized in European hospitals and healthcare systems. *BMC Health Services Research*, 14, 478.

Wiig, S., Guise, V., Anderson, J., Storm, M., Lunde Husebø, A.M., Testad, I., Søyland, E. and Moltu, K. (2014c). Safer@home – Simulation & training: The study protocol of a qualitative action research design. *BMJ OPEN*, 4, e004995. doi: 10.1136/bmjopen-2014-004995.

Wiig, S., Robert, G., Anderson, J., Pietikainen, E., Reiman, T., Macchi, L. and Aase, K. (2014a). Applying different quality and safety models in healthcare improvement work: Boundary objects and system thinking. *Reliability Engineering & System Safety*, 125, 134–144. doi: 10.1016/j.ress.2014.01.008.

Wiig, S., Storm, M., Aase, K. et al. (2013). Investigating the use of patient involvement and patient experiences in quality improvement in Norway: Reality or rhetoric? *BMC Health Services Research*, 13, 206.

Yin, R.K. (ed.). (2004). *The Case Study Anthology.* Thousand Oaks, CA: Sage Publications.

4 What Is the Role of Theory in Research on Patient Safety and Quality Improvement? *Is There a Nordic Perspective to Theorising?*

Karina Aase and Jeffrey Braithwaite

CONTENTS

In this chapter, we will discuss ways in which theory plays a role in research on patient safety and quality improvement. In general, theory provides us with the capacity to understand and account for empirical phenomena, to explain relationships between constructs and to predict conditions under which relationships are more likely to hold. More specifically, in improvement and implementation work and other applied domains, theory gives us a framework by which to conduct evaluations to understand what works and why. Within research on patient safety and quality improvement in healthcare, the role of theory has so far been under-reported or under-recognised, or both. We will investigate the reasons for this and look at different 'applications' of theory within patient safety and quality improvement, respectively. We will also reflect on the question of whether there is a possible – or potential – Nordic perspective to theorising within patient safety and quality.

INTRODUCTION

In order to embark on tackling these issues, it is useful to consider whether the Nordic countries as a collective might in fact be different from other parts of the world. Responding to this question feeds into our overall quest to appreciate Nordic patient safety and quality improvement and Nordic theorising. Our immediate answer is yes and no.

Supporting the no side, we note some obvious facts about Western society – that countries which comprise 'the West' are increasingly homogeneous, with converging cultures, economies and world views. We may characterise this phenomenon by its technical term – globalisation, also called the McDonaldization (Ritzer, 1993) of the world. And yet there are obvious deep cultural, literary and societal differences which set countries and regions apart. The Nordic countries have a rich history which stamps their contemporary distinctiveness (Hermannsson, 1910; Hollander, 1997; Christiansen, 2006). Threads of history and literature feed into the modern manifestation of Nordic distinctiveness including social solidarity, welfareism, individual hardiness and a humane outlook on life. For a further discussion of aspects of Nordic (and international) distinctiveness and cultural characteristics, see House et al. (2004). From House et al.'s GLOBE study of comparative international differences, we know that Nordic European societal culture has features such as an orientation towards performance and a humane orientation. It also has features, according to this research, of relatively low levels of assertiveness, both in values and in practice. These features are relative and characterise Nordic society compared to, for example, U.S. individualism.

WHAT DO WE MEAN BY THEORY?

Let us now assess theory and its use in patient safety and quality improvement. Theory can mean anything from a simple guess or mere speculation to articulation of a set of interrelated and logical propositional constructs that provide explanatory and predictive power (Mark et al., 2004). The diverse meanings that can be attributed to the word 'theory' led Merton (1967) to suggest that it has the potential to obscure rather than create understanding. Despite this, the development of theory that leads to an understanding of patient safety and quality improvement should be of paramount importance for several reasons (West, 2001; Walshe, 2007). Theory can be used, for example,

1. As a framing device to call attention to some features of the phenomenon being studied and to direct attention away from other aspects
2. To aggregate knowledge such that it can then be generalised from one setting to another
3. To recommend ways of organising to improve safe and high-quality patient care
4. To avoid simplistic or standardised solutions or interventions to complex patient safety and quality improvement initiatives for which high levels of variance in context, content and application are often inherent and desired characteristics

A common description of the elements in a theory in healthcare is provided by Meleis (1997):

> an organized, heuristic, coherent and systematic articulation of a set of statements related to significant questions that are communicated in a meaningful whole for the purpose of providing a generalizable form of understanding.

Notwithstanding that we can arrive at a comprehensive definition of this type, different disciplines sometimes have diverging views of what constitutes theory, and there are many kinds of theory. There is a diverse range of theories potentially relevant to patient safety and quality improvement, which encompass a wide variety of disciplines from anthropology, psychology, sociology, behavioural economics and management and organisational sciences. Meleis' definition of theory can be characterised as constituting a description of a formal, publicly developed theory (Davidoff et al., 2015). Alongside formal theories are the more informal experience-based theories most often at work in improvement efforts. The challenge according to Davidoff and colleagues is to combine the two theory domains.

Because formulating 'good theory' is inherently difficult, Sutton and Staw (1995) identified examples of what theory is not. According to them, theory is not represented by an extensive list of references in which logical relationships are explained elsewhere. Models and variables must be accompanied by logical statements that explain not only how variables are related but also why they are related. Similarly, theory is not represented by a mere description of observed relationships, beta weights or factor loadings. Data can describe relationships between variables but cannot explain why the relationships hold.

Nor is theory a laundry list of variables or a diagram used to depict a structural equation model. And nor is theory rendered – at least not explicitly – in evidence-based clinical practice guidelines or in so-called best practices. Theory results from a process of systematic and logical reasoning, either deductive or inductive, through which the occurrence or nonoccurrence of some phenomenon can be understood and predicted (Sutton and Staw, 1995, p. 11 in Mark et al., 2004).

Davidoff et al. (2015) following Merton (1968) make a useful distinction between *grand theory*, *mid-range theory* and *programme theory*. Grand theory is formulated at a high level of abstraction, making generalisations that apply across many different domains (e.g. a theory of social inequality). Mid-range (or middle-range) theories are delimited in their area of application and are intermediate between 'working hypotheses' and a 'conceptual scheme' (e.g. the theory of diffusion of innovations). Programme theories provide a 'small theory' for each intervention being purposefully practical and accessible (e.g. working models, logic model, driver diagram).

THE 'UNDER-USE' OF THEORIES AND THE DEVELOPMENT OF A FRAMEWORK

Many healthcare professionals, including practitioners responsible for quality improvement, are mystified and often alienated by the word 'theory' and by theories themselves, which because they can be complex, or abstract, or both, discourages them from using them in their work and research (Davidoff et al., 2015). The potential 'users' of theory furthermore form a diverse group, ranging from improvers (who might be practitioners, managers or policymakers) to academic researchers. The 'improvers' are interested in theory to the extent that it can help them do their work better, if they want to have or apply theory at all. For the 'researchers', theory itself is frequently the object of study with the aim to confirm, disconfirm or refine an extant theory – or build a new one. It is in part the differences in the interests and practices of improvers and researchers that explain the underlying distinctions between improvement projects and research studies (Davidoff, 2014). Thus, both groups differ in their underlying epistemologies and secure and apply knowledge in different ways.

Theory has not commonly been used in the fields of patient safety and quality improvement research (Grol et al., 2007; Walshe, 2007; Foy et al., 2011; Lipworth et al., 2013) – at least, explicitly. A review of 235 implementation studies found that only 53 mentioned or used theory in any demonstrable way and that only 14 were explicitly theory based (Davies et al., 2010). Similarly, most reports of patient safety practice evaluations do not report any theoretical model underpinning the intervention. Even for the most commonly studied patient safety practices such as checklists, medication reconciliation and falls prevention, a review of published evaluations found only two studies that even partially reported a theory for why the practices should work (Foy et al., 2011). Whether or not researchers *report* their use of theory or their hypotheses, it is nevertheless obvious we *use* theories and hypotheses all the time. Everyone formulates and is driven by theories and hypotheses but they remain relatively invisible. Policymakers issue a policy, theorising that this will shape others' behaviours; clinicians treat patients, theorising that they will get well as a consequence; patients trust their clinicians, theorising that this is a better bet than the alternative.

Why might it be the case, then, that researchers and improvement practitioners within healthcare quality and safety are so poor at explicitly mobilising theory to support their findings or interventions? Lipworth et al. (2013) point to several reasons:

1. Unawareness of relevant theories or neglect to draw upon any conceptual perspective.
2. Difficulty in choosing from the abundance of theories available.
3. If applying a particular theory, the one selected may prove insufficiently broad to account for all relevant factors in the field of study.

In an attempt to solve some of these issues, the theoretical domains framework (TDF) has been designed to apply in a variety of settings (Michie et al., 2005). The TDF consists of a set of conceptual determinants and associated constructs from psychological and organisational theory that is assumed to influence behaviour and behavioural change. The framework was established to more accurately identify barriers and levers and to design interventions with sufficient theoretical richness to address them (Michie et al., 2008). The TDF consists of 14 domains: knowledge, skills, social and professional role and identity, beliefs about capabilities, optimism, beliefs about consequences, reinforcement, intention, goals, memory, attention and decision processes, environmental context and resources, social influences, emotion and action planning (for more details, see Cane et al., 2012).

In a thematic synthesis of 50 studies of various clinical quality interventions mapped to the domains of the TDF, Lipworth et al. (2013) found that the framework could be used to map most, if not all, of the attitudinal and behavioural barriers and facilitators of uptake of clinical quality interventions. All 14 domains emerged as relevant, and the authors suggested a possible additional domain, taking the total to 15, related to perceived trustworthiness of those instituting clinical quality interventions.

SAFETY SCIENCE THEORIES AND THEIR APPLICATION IN HEALTHCARE

The safety science research field in general, at least among some adherents, has been occupied, and sometimes preoccupied, with theorising. Anthropologists, sociologists and critical realists have, for example, discussed and created models for understanding risk (Parker and Stanworth, 2005), while others have created theories for understanding the organisational and cultural drivers for safety (e.g. Braithwaite et al., 2010). In the following, we have listed six different theoretical perspectives manifesting in the safety science field (Aase and Rosness, 2013):

1. The *energy and barrier perspective*, according to which accidents can be understood and prevented by focusing on dangerous energies and means by which such energies can reliably be separated from vulnerable targets (Gibson, 1961; Haddon, 1970, 1980). Reason's Swiss Cheese Model is probably the most well-known example of such defence-in-depth barrier thinking, also applied frequently within healthcare (Reason, 2000).

2. Perrow's theory of *normal accidents*, which explains some major accidents in terms of a mismatch between the properties of the technology to be controlled and the structure of the organisation responsible for controlling the technology (Perrow, 1984).
3. The theory of *High Reliability Organisations* (HRO), which proposes that some organisations have achieved an outstanding capacity to handle complex technologies without generating major accidents by building organisational redundancy and a capacity to reconfigure in adaptation to peak demands and crisis (Rochlin et al., 1987; LaPorte and Consolini, 1991).
4. *The information processing perspective* (Turner, 1978; Turner and Pidgeon, 1997), which conceptualises an accident as a breakdown in the flow and interpretation of information that is linked to physical events.
5. The *decision-making perspective*, with a focus on the handling of conflicting objectives, including accounts of how activities migrating towards the boundary of acceptable performance.
6. The *resilience engineering* perspective, which combines and elaborates concepts and ideas from previous perspectives in an effort to build a more coherent understanding of resilience in socio-technical systems and to provide tools to help organisations monitor and build resilience (Hollnagel et al., 2006).

These different theoretical perspectives have distinguishable foci and operate at different levels. Despite their increasing popularity in the broader safety science field, they have had limited applications within the patient safety and quality improvement domains in healthcare, with some exceptions we describe below.

Root Cause Analysis, Swiss Cheese Model (Energy and Barrier Perspective)

Root cause analysis (RCA) is now widely used in many healthcare systems internationally. It stems from James Reasons' conceptualisation of human error based on a psychological tradition whereby errors are explained by linear cause–effect chains and prevented by a set of barriers in the so-called defence-in-depth (Reason, 1990). Following from this, the well-known Swiss Cheese Model was developed, visualising the defence-in-depth principle. RCA is a method of identifying the root causes of faults or adverse events based on a linear cause–effect model. The root cause prevents the final undesirable event from recurring, whereas a causal factor is one that affects an event's outcome but is not a root cause (see Braithwaite et al., 2006). RCA is one of the most widely used methods to improving patient safety, but few data exist that uphold its effectiveness, and critics are abundant (e.g. Rasmussen, 1997; Braithwaite et al., 2015). The quality of RCA varies across facilities, and its effectiveness in lowering risk, or improving patient safety, has not been systematically established (e.g. Iedema, 2006; Shaqdan et al., 2014).

Following from the focus on RCA, the patient safety research field has also been occupied with the taxonomy or classification of patient safety concepts. The World Health Organization's World Alliance for Patient Safety has developed an International Classification for Patient Safety. This categorisation system aims to transform patient safety information collected from disparate systems into a common format, thereby facilitating aggregation, analysis and learning. The conceptual framework consists of 10 high level classes: incident type, patient outcomes, patient characteristics, incident characteristics, contributing factors/hazards, organisational outcomes, detection, mitigating factors, ameliorating actions and actions taken to reduce risk. The 10 classes are presented in a flow chart (Sherman et al., 2009). The basic principle behind the theory involves clarifying predefined characteristics in order to manage future incidents.

Learning and Reporting (HRO Perspective)

Partly in line with the taxonomic view of errors and adverse events reported earlier, learning and feedback from past experience is seen as a major building block in preventive safety work. The 2000 Institute of Medicine (IOM) report (Kohn et al., 2000) identified under-reporting of adverse events as a patient safety issue and recommended that hospitals develop non-punitive environments to promote incident reporting. Patient safety advocates called for intensifying the reporting and analysis of adverse events and near-misses, and most hospitals have today implemented a variety of reporting systems modelled, in part, on experiences in the aviation industry (Tamuz and Harrison, 2006).

The hallmark report 'An organisation with a memory' (UK Department of Health, 2000) supported the arguments raised earlier but more clearly related the issues of learning from adverse events to the theoretical areas of organisational culture and safety information systems. Currently, multiple countries operate with mandatory national systems for reporting and learning from adverse events. Denmark became the first country to introduce nationwide mandatory reporting, as laid out in the Danish Act on Patient Safety in 2004. The Act obligates frontline personnel to report adverse events to a national reporting system. Hospital owners are obligated to act on the reports, and the National Board of Health is obligated to communicate the learning nationally. The reporting system is intended for learning and frontline personnel cannot be sanctioned for reporting (Danish National Board of Health, 2007).

Resilient Healthcare (Resilience Engineering Perspective)

Applying resilience engineering concepts to healthcare has become an increasingly important research area within patient safety circles over the last decade. In a review of the resilience engineering literature assessing 237 studies, Ringhi et al. (2015) found that healthcare accounted for 19% of the studies. They also reported that theoretical discussions or contributions accounted for more than 50% of the resilience engineering studies (including 'why resilience?' and 'what is resilience?').

The major thread of the argument of resilient healthcare theory is to appreciate that healthcare is resilient to a large extent and that everyday performance, such as clinical practice, succeeds more often than it fails. Healthcare personnel adjust their practices to match the conditions. Facilitating work flexibility, and actively trying to increase the capacity of healthcare personnel to deliver care more effectively, is a key component in resilient healthcare. This involves focusing on how everyday performance usually succeeds rather than on why it occasionally fails and actively working towards improving the former rather than preventing the latter (Hollnagel et al., 2013; Wears et al., 2015; Braithwaite et al., in press).

QUALITY IMPROVEMENT THEORIES AND THEIR APPLICATION IN HEALTHCARE

The quality improvement literature is large and diverse. Theories and models are not always well defined and healthcare improvers, researchers and organisations often draw on a range of tools and principles from different approaches (Powell et al., 2009). Unlike parts of the safety science field preoccupied with theoretical debates, research on quality in healthcare has been more empirically driven, focusing on practical, technical tools for improvement. However, there is also growing awareness that a solely technical–empirical approach to quality improvement will not be sufficient to embed and sustain the organisational change necessary to improve quality. Organisational, cultural and other contextual factors are increasingly considered crucial to the success of quality improvement initiatives (Powell et al., 2009; Wiig et al., 2014).

Another point where quality improvement and patient safety research departs from the broader safety science field is in the degree of applications of different theoretical perspectives or models in healthcare settings. As indicated earlier, we have seen limited application of safety theories and models in healthcare, with some exceptions. Within the quality improvement field more broadly there are abundant applications of a diverse set of theories and models in healthcare. Most of these theoretically based quality improvement approaches (with the marked exception of Donabedian's (1988) structure–process–outcome framework) come originally from manufacturing industries in the field of total quality management and continuous quality improvement processes and techniques (Berwick, 1996). Later, we will briefly mention some of the approaches applied in healthcare. The descriptions are based on Sales (2013), Powell et al. (2009), Boaden (2009) and Boaden et al. (2008).

Structure–Process–Outcome Framework

Donabedian's (1988) framework for understanding factors influencing the quality of care, and in particular outcomes of health services, has been influential in practice and widely adopted in the literature on healthcare quality improvement. Donabedian proposed that the *structure* of health services (e.g. physical facilities, types of services available, staffing levels) influences the *process* of care. Care processes include specific interventions such as surgery, prescribing medications and monitoring processes of care. These factors influence *outcomes*, which can be at several different levels, although the original framework focuses on patient-level outcomes of care. These outcomes include whether a patient survives a care episode delivered during an acute event, the quality of life someone has after receiving health services and other sequelae of the health condition and of health services received (e.g. adverse events). In later work, Donabedian also focused on system-level outcomes, such as cost and efficiency (e.g. 1990). Widely adopted, but also critiqued for being too simple and overtly linear, the structure–process–outcome framework was designed (in Donabedian's own words) as 'a handy classification scheme'. It has been influential, too, as an input in the development of other approaches, for example, the seven pillars for quality (similar to the six quality principles outlined by the United States's IOM) and the 11 essential principles for quality assurance (Schiff and Rucker, 2001).

Plan–Do–Study–Act Cycle

The Institute for Healthcare Improvement (IHI) in the United States has been the most prominent advocate for the application of continuous quality improvement in healthcare. The tenets of quality improvement according to this approach include the use of data and statistical analysis to identify and control processes; the use of benchmarking; the use of teams to identify problems, processes and solutions and the use of some form of improvement process, usually described as a cycle: *plan*, *do*, *study* (or *check*) and *act*. The Plan–Do–Study–Act (PDSA) cycle was originally developed by Deming (1986) and presented as the basis of the 'Model for Improvement' (Langley et al., 1996). The approach is typically applied repeatedly in small cycles and is sometimes suggested as one of the phases in Six Sigma and Lean thinking. IHI has developed the principles into processes called *collaboratives* involving a number of teams working together with a group of experts to plan, implement and monitor improvements in care.

The impact of PDSA cycles accompanied with collaborative arrangements in healthcare appears to depend on the focus of the collaborative, the participants and their host organisation and the style and method of implementing the collaborative (Øvretveit et al., 2002). Its application in healthcare has been studied (compared with other approaches), although there is little evidence to suggest it is more cost-effective than any other approach. There is mixed evidence of its efficacy and contributions to improvement (Taylor et al., 2014).

Six Sigma

Six Sigma is an improvement or redesign approach developed initially in manufacturing. The approach has been heavily promoted in the U.S. healthcare system. It is presented as a systematic method for strategic process improvement and development relying on statistical methods to reduce deficiencies that customers can identify, also known as 'customer-defined defect rates' (Linderman et al., 2003). A common method within Six Sigma includes the process improvement methodology similar to the PDSA cycle: Define, Measure, Analyse, Improve and Control. Many organisations appear to have re-labelled total quality management as Six Sigma, and a recent development is the use of 'Lean Six Sigma', meant to facilitate streamlining of processes.

There are a variety of reports on the application of Six Sigma in healthcare but few that take a systematic approach to assessing its effectiveness. The evidence is descriptive, with little by way of fundamental critique, examination of effectiveness or independent evaluation.

Statistical Process Control

Statistical Process Control (SPC) methods can be traced to Shewhart (1931) who identified the difference between 'natural' variation in measures of a process (termed the 'common cause') and variation that could be controlled (termed the 'assignable' cause). Processes that exhibited only common cause variation are said to be in a state of statistical control. SPC is a key tool in the total quality management framework and is also used in Six Sigma and sometimes with the PDSA cycle. Recognition of the importance of variation in healthcare has led to an interest in SPC, and a variety of applications have been reported (Thor et al., 2007). Its application may be limited by the extent to which the objective of improvement is the reduction of variation, the complexity and appropriateness of data sets representing aggregations of different types of patients or units and the implications of the underlying statistics of having very small or very large data sets.

System-Wide Multi-Model Approaches

In practice, quality improvement models and their tools are used in a variety of ways. They are rarely applied singly or sequentially; what is more common in healthcare settings is to draw on combinations or hybrids of the main approaches. Powell et al. (2009) have reviewed healthcare settings that stand out in the literature as particularly successful or integrated examples of system-wide approaches to quality improvement. The Jönköping quality improvement programme (Sweden; see description later in the chapter), the Kaiser Permanente system (United States), the Quality Enhancement Research Initiative of the Veterans Health Administration (United States) and the Organising for Quality leading hospitals (United Kingdom, the Netherlands, United States) are all examples of such system-wide approaches to quality improvement based on a variety of theoretical approaches and models. While each is a unique example, they have common elements in their programmes: effective leadership, an emphasis on teamwork, receptive organisation cultures and support for continuous improvement of care processes. And what stands out in the description and evaluation of these blended approaches is that quality improvement models need to be supported with networks, structures, cultures and systems to undergird the 'organisational and human processes' of quality.

Quality Improvement Healthcare Approaches Summarised

Reviewing the models and system-wide approaches to quality improvement shows that there are strong commonalities between them: although they may have different emphases, many share similar theoretical constructions, underlying objectives and the distinctions between the approaches are often blurred in practice (Powell et al., 2009). Moreover, each of the approaches and the data used to underpin them can be used either to *enable* quality improvement by 'inspiring and developing' or to *mandate* quality improvement through 'policing, punishing and rewarding'.

Despite the many insights into implementation emerging from the literature, it remains challenging to assess the impact of specific theories, models or approaches or to make comparisons across applications. There does seem to be a consensus that there is no one 'right' or preferred quality improvement theory, framework or model. Instead, successful quality improvement is more about the interaction between any given approach and its implementation in the local context (Ovretveit et al., 2011; Mittman, 2015).

A NORDIC PERSPECTIVE?

So, explicit applications of theory seem to be under-represented or under-reported in the international literature on patient safety and quality (Walshe, 2007; Davies et al., 2010; Foy et al., 2011), but many theories, frameworks and models are used in practice. As documented elsewhere in this book, there seems to be no exception to this general rule in the case of Nordic peer-reviewed academic journal articles within quality and patient safety between 2000 and 2014. A main finding in the literature review reported in Chapter 1 was the lack of theory in use. Of 163 Nordic journal articles only 8 (Johansen et al., 2004; Øvretveit, 2007; Rahimi et al., 2009; Hovlid et al., 2012; Ingemansson et al., 2012; Jakobsson and Wann-Hansson, 2013; Hovlid and Bukve, 2014; Jakobsen et al., 2014) mentioned a theoretical perspective in the paper's abstract (e.g. implementation theory, model for understanding quality implementation, organisational learning).

One might deduce from this that there is no Nordic flavour to theorising within quality and patient safety. Again our answer to the question is yes and no. Despite the fact that, mirroring the broader literature, explicit theorising seems to be missing in Nordic peer-reviewed articles over the last 15 years, we can nevertheless discern that selected Nordic scholars in the past have put a substantial stamp of distinctiveness on their work. Nordic scholars have developed or used theories and influenced the theoretical thinking of others within patient safety and quality improvement activities. And further, contemporary Nordic researchers seem to bear the hallmark of these earlier influences.

We can trace some of this Nordic influence by drawing on the work of three scholars. These are Jens Rasmussen, Erik Hollnagel and John Øvretveit, and a setting – Jönköping, in Sweden. Within the safety science field, the well-known Danish scholar Rasmussen provided key concepts for understanding safety and accidents, many of which resonate through the decades and remain relevant today. His publications, spanning from 1969 to 2000, are too early to be included in the more recent literature review presented in Chapter 1. Yet, Nordic (and international) authors with different disciplinary influences (e.g. psychology, management and sociology) and orientations in the field of safety have indeed incorporated, in different ways, Rasmussen's ideas into their studies, building upon different aspects of his research over several decades. Principles such as degree of freedom, self-organisation and adaptation, the defence-in-depth fallacy and also Rasmussen's notions of error still offer powerful insights into the challenge of predicting and preventing major accidents (LeCoze, 2014). Even though Rasmussen did not specifically direct the majority of his work towards healthcare, many of his ideas are prevalent in the patient safety research field. Especially, Rasmussen's move from a micro view of accidents to a macro (socio-technical perspective), one in which he was trying to apprehend the bigger picture of error, and his views of where safety science is seen as a cross-disciplinary pursuit

(Rasmussen, 1997, 2000; Rasmussen and Svedung, 2000) have sparked some of the current trends in healthcare safety (e.g. Nemeth et al., 2008; Hollnagel et al., 2013).

One of Rasmussen's early collaborators and later a critic of among others his 'skill-rule-knowledge' model, Erik Hollnagel is another highly influential Danish safety science scholar. Hollnagel has in many ways elaborated on some of Rasmussen's ideas related to self-organisation and the principles of a functional versus structural approach. He is one of the main thinkers behind resilience engineering and later resilient healthcare theories (e.g. Hollnagel et al., 2013; Hollnagel, 2014). As such, Hollnagel has come to be seen as a leading scholar and invaluable knowledge broker, bridging the divide between safety science and patient safety, for example. Hollnagel's work on theorising healthcare resilience has mainly been published in books and is therefore not included in the literature review referred to in Chapter 1. It is nevertheless among the most persuasive of recent work, and one which we predict will become increasingly influential.

Within the Nordic quality improvement research field, the Jönköping Quality Program is widely known in Sweden and is seen internationally as a long-running example of a successful system-wide improvement programme (Bodenheimer et al., 2007; Øvretveit and Staines, 2007; Staines et al., 2015). Jönköping County Council has gone further than most health systems in building an impressive infrastructure for quality improvement and of helping make learning and improvement part of everyday clinical practice. The emphasis on a 'bottom-up' approach, and refraining from 'forcing' individuals or departments into quality improvement, has been characteristic. The Swedish culture and conditions allow this in ways that may be more difficult in other contexts (Øvretveit and Staines, 2007). In a later evaluation of the Jönköping Quality Program, Staines and colleagues (2015) found five key issues shaping the improvement programme over the last 10 years:

1. A rigorously managed succession of the chief executive officer
2. Adept management of a changing external context
3. Clear strategic direction relating to integration
4. A broadened conceptualisation of 'quality' (incorporating clinical effectiveness, patient safety and patient experience)
5. Continuing investment in quality improvement education and research

As the Jönköping example indicates, we may claim that theorising in quality improvement research takes a different and more applied approach than it does in patient safety research. Implementation research has been closely linked to quality improvement. In this field, the Swedish scholar John Øvretveit has played a key role in evaluating improvement and implementation efforts, influencing the Nordic quality and safety research agenda. Øvretveit has for decades advocated for evidence- and theory-based evaluation of both quality improvement and patient safety (e.g. Øvretveit, 2014). Fifteen years ago, Øvretveit published work on the Norwegian approach to integrated quality development (Øvretveit, 2001) and was also influential in the evaluation of the Swedish Jönköping Quality Program (Øvretveit and Staines, 2007). Øvretveit has been a strong advocate for the development and use of programme theory, that is intervention-specific working models or hypotheses (e.g. Foy et al., 2011; Øvretveit, 2014).

As shown in Chapter 1, articulating a patient perspective in quality improvement and patient safety seems prominent in the Nordic literature. There is solid extant documentation on the evaluation, experiences and perceptions of healthcare services from the perspective of patients and next of kin. Even though no specific theoretical foundation exists for the extensive Nordic incorporation of the patient perspective into Nordic improvement efforts, we nevertheless suggest this is a cornerstone of a Nordic health model and approach to healthcare provision. This favours, and entails, principles of equality, the legal rights of patients and minimal power distance (Norden, 2013, 2014) between providers and recipients of care. This, we further suggest, is part of a humane, inclusive approach rooted in societal solidarity and welfareism, which we identified earlier, and which characterised Nordic thinking and theorising.

IMPLICATIONS – 'GOOD' THEORISING WITHIN PATIENT SAFETY AND QUALITY IMPROVEMENT RESEARCH?

Efforts to build theory that is useful for both research and practice will depend on the extent to which we are able to identify constructs that capture the complex interface between the clinical, behavioural and organisational domains within which healthcare personnel and patients coexist and collaborate (Mark et al., 2004). Several such frameworks have emerged such as the TDF (Michie et al., 2005, 2008; Cane et al., 2012), theory-driven improvement evaluations (Grol et al., 2007; Walshe, 2007; Foy et al., 2011; Davidoff et al., 2015) and frameworks for building theories of context (Kaplan et al., 2012; McDonald, 2013). To a considerable degree, theories, models and frameworks partly overlap and several of them underpin quality improvement interventions. Within safety science and more specifically within patient safety, there is a multiplicity of theoretical perspectives, and they appear in fragmented and piecemeal ways. No universal frameworks exist, but nor would we expect this. Pronovost et al. (2009), for example, developed a framework for patient safety research and improvement (see also Chapter 1) to clarify broad clinical and policy domains that link to safety. The taxonomy is presented as a prioritisation tool rather than a theoretical framework.

Synthesising what we have said in this chapter, then, we present a set of generic issues that improvers and researchers might have regard to when theorising within patient safety and quality, regardless of research topic or domain:

- Seek to make explicit the theoretical assumptions behind the choice of research focus or improvement efforts (Grol et al., 2007; Davidoff et al., 2015).
- Describe contextual factors (e.g. external, organisational, professional, managerial, cultural, implementation–orientation) that might have an impact on the theoretical assumptions of any research topic or improvement efforts (Foy et al., 2011; Kaplan, 2012; Mc Donald, 2013).

- Base any initial theoretical assumptions on literature reviews, consensus-driven approaches and/or conceptual frameworks (Foy et al., 2011; Davidoff et al., 2015).
- Base any theorising on interdisciplinary collaborations (e.g. medical, nursing, psychology, sociology, management science) that can identify the core of the theoretical perspectives (Foy et al., 2011).
- Be alert to the possibility of being led astray by apparently attractive theories that may be partially inappropriate for the context or are flawed (Davidoff et al., 2015).

CONCLUSION

Theory takes many forms, some informal, some highly structured. Formal theory can complement informal, experience-based theory, helping to define areas of dysfunction in patient safety and quality, pinpointing their loci and identifying their possible mechanisms (Davidoff et al., 2015). Formal theoretical frameworks can allow the accumulation of knowledge based on informal or small theories and empirical observations.

All in all, in this chapter we have argued for the importance of theory for improvers and researchers, and we would emphasise by way of concluding, four key points.

1. Explicit and formalised theorising is largely missing in research and practice within quality and safety in healthcare.
2. When theorising does occur, it takes on very different forms within the fields of quality and safety.
3. There are many theoretical perspectives, theory-oriented models and frameworks on which to draw.
4. We can trace the contribution of early theories, theorists and scholars such as Rasmussen, Hollnagel and Øvretveit and a setting, Jönköping, and see their ideas and influence having a discernible Nordic flavour, and as being reflected in current Nordic and other international improvement work and safety and quality research.

REFERENCES

Berwick, D.M. (1996). Harvesting knowledge from improvement. *Journal of the American Medical Association*, 275(11), 877–878.

Boaden, R. (2009). Quality improvement: Theory and practice. *British Journal of Healthcare Management*, 15(1), 12–16.

Boaden, R., Harvey, G., Proudlove, N. and Moxham, C. (2008). *Quality Improvement in Healthcare: Theory and Practice*. Coventry, U.K.: NHS Institute for Innovation and Improvement/Manchester Business School.

Bodenheimer, T., Bojestig, M. and Henriks, G. (2007). Making system-wide improvements in health care: Lessons from Jönköping County, Sweden. *Quality Management in Health Care*, 16(1), 10–15.

Braithwaite, J., Hyde, P. and Pope, C. (eds.). (2010). *Culture and Climate in Health Care Organizations*. London, U.K.: Palgrave Macmillan.

Braithwaite, J., Wears, R.L. and Hollnagel, E. (2015). Resilient health care: Turning patient safety on its head. *International Journal for Quality in Health Care*, 27(5), 1–3. doi: 10.1093/intqhc/mzv063.

Braithwaite, J., Wears, R.L. and Hollnagel, E. (eds.). (in press). *Reconciling Work-as-Imagined and Work-as-Done*. Farnham, U.K.: Ashgate Publishing Ltd.

Braithwaite, J., Westbrook, M., Mallock, N. and Travaglia, J. (2006). Experiences of health professionals who conducted root cause analyses after undergoing a safety improvement programme. *Quality & Safety in Health Care*, 15(6), 393–399.

Cane, J., O'Connor, D. and Michie, S. (2012). Validation of the theoretical domains framework for use in behaviour change and implementation research. *Implementation Science*, 7(37), 1–17.

Christiansen, E. (2006). *The Norsemen in the Viking Age*. New York: John Wiley & Sons Ltd.

Davidoff, F., Dixon-Woods, M., Leviton, L. and Michie, S. (2015). Demystifying theory and its use in improvement. *BMJ Quality and Safety*, 24, 228–238. doi: 10.1136/bmjqs-2014-003627.

Davies, P., Walker, A.E. and Grimshaw, J.M. (2010). A systematic review of the use of theory in the design of guideline dissemination and implementation strategies and interpretation of the results of rigorous evaluations. *Implementation Science*, 5(14), 1–6. doi: 10.1186/1748-5908-5-14.

Deming, W.E. (1986). *Out of the Crisis*. Cambridge, MA: The MIT Press.

Donabedian, A. (1988). Monitoring: The eyes and ears of healthcare. *Health Progress*, 69(9), 38–43.

Donabedian, A. (1990). Quality and cost: Choices and responsibilities. *Journal of Occupational Medicine*, 32(12), 1167–1172.

Foy, R., Ovretveit, J., Shekelle, P.G., Taylor, S.L., Dy, S. Hempel, S., McDonald, K.M., Rubenstein, L.V. and Wachter, R.M. (2011). The role of theory in research to develop and evaluate the implementation of patient safety practices. *BMJ Quality and Safety*, 20, 453–459. doi: 10.1136/bmjqs.2010.047993.

Grol, R.P.T.M., Bosch, M.C., Hulscher, M.E.J.L., Eccles, M.P. and Wensing, M. (2007). Planning and studying improvement in patient care: The use of theoretical perspectives. *The Milbank Quarterly*, 85(1), 93–138.

Hermannsson, H. (1910). *Bibliography of the Sagas of the Kings of Norway and Related Sagas and Tales*. Ithaca, NY: Cornell University Library. Online at https://archive.org/details/bibliographyofsa03hermuoft (accessed 3 September 2015).

Hollander, L.M. (1997). *Njal's Saga*. Hertfordshire, U.K.: Wordsworth Editions.

Hollnagel, E. (2014). *Safety-I and Safety-II: The Past and Future of Safety Management*. Farnham, U.K.: Ashgate Publishing Ltd.

Hollnagel, E., Braithwaite, J. and Wears, R.L. (eds.). (2013). *Resilient Health Care*. Farnham, U.K.: Ashgate Publishing Ltd.

House, R.J., Hanges, P.K., Javidan, M., Dorfman, P.W. and Gupta, V. (eds.). (2004). *Culture, Leadership, and Organizations*. London, U.K.: Sage Publications.

Iedema, R.A.M., Jorm, C., Long, D., Braithwaite, J., Travaglia, J. and Westbrook, M. (2006). Turning the medical gaze in upon itself: Root cause analysis and the investigation of clinical error. *Social Science & Medicine*, 62, 1605–1615.

Kaplan, H.C., Provost, L.P., Froehle, C.M. and Margolis, P.A. (2012). The Model for Understanding Success in Quality (MUSIQ): Building a theory of context in healthcare quality improvement. *BMJ Quality & Safety*, 21(1), 13–20.

Langley, G.J., Nolan, K.M., Norman, C.L., Provost, L.P. and Nolan, T.W. (1996). *The Improvement Guide: A Practical Approach to Enhancing Organizational Performance*. San Francisco, CA: Jossey Bass.

Linderman, K., Schroeder, R.G., Zaheer, S. and Choo, A.S. (2003). Six Sigma: A goal-theoretic perspective. *Journal of Operations Management*, 21(2), 193–203.

Lipworth, W., Taylor, N. and Braithwaite, J. (2013). Can the theoretical domains framework account for the implementation of clinical quality interventions? *BMC Health Services Research*, 13, 530.

Mark, B., Hughes, L.C. and Jones, C.B. (2004). The role of theory in improving patient safety and quality health care. *Nursing Outlook*, 52, 11–16. doi: 10.1016/j.outlook.2003.10.010.

McDonald, K.M. (2013). Considering context in quality improvement interventions and implementation: Concepts, frameworks, and application. *Academic Paediatrics*, 13, 45.

McLaughlin, C.P. and Kaluzny, A.D. (2004). *Continuous Quality Improvement in Health Care: Theory, Implementation, and Applications*, 2nd edn. Sudbury, MA: Jones and Bartlett.

Meleis, A.I. (1997) *Theoretical Nursing. Development and Progress*, 3rd edn. New York: Lippincott.

Merton, R. (1967). *On Theoretical Sociology*. New York: Free Press.

Merton, R.K. (1968). *Social Theory and Social Structure*. New York: Free Press.

Michie, S., Hardeman, W., Johnston, M., Francis, J. and Eccles, M. (2008). From theory to intervention: Mapping theoretically derived behavioural determinants to behaviour change techniques. *Applied Psychology*, 57, 660–680.

Michie, S., Johnston, M., Abraham, C., Lawton, R., Parker, D. and Walker, A. (2005). Making psychological theory useful for implementing evidence based practice: A consensus approach. *Quality and Safety in Health Care*, 14, 26–33.

Mittman, B. (2015). Implementation science – The next frontier. In Barach, P.R., Jacobs, J.P., Lipshultz, S.E. and Laussen, P.C. (eds.), *Pediatric and Congenital Cardiac Care*, Vol. 2: Quality Improvement and Patient Safety. London, U.K.: Springer-Verlag, pp. 285–292.

Nemeth, C., Wears, R.L., Woods, D., Hollnagel, E., and Cook, R. (2008). Minding the gaps: Creating resilience in health care. In Henriksen, K., Battles, J.B., Keyes, M.A. et al. (eds.), *Advances in Patient Safety: New Directions and Alternative Approaches*, Vol. 3: Performance and Tools. Rockville, MD: Agency for Healthcare Research and Quality, pp. 1–13 (online version).

Øvretveit, J. (2001). The Norwegian approach to integrated quality development. *Journal of Management in Medicine*, 15(2), 125–141.

Øvretveit, J. (2014). *Evaluating Improvement and Implementation for Health*. Milton Keynes, U.K.: Open University Press, McGraw-Hill Education.

Øvretveit, J., Bate, P., Cleary, P. et al. (2002). Quality collaboratives: Lessons from research. *Quality and Safety in Health Care*, 11(4), 345–351.

Øvretveit, J., Shekelle, P., Dy, S., McDonald, K., Hemper, S., Pronovost, P., Rubenstein, L., Taylor, S., Foy, R. and Wachter, R. (2011). How does context affect interventions to improve patient safety? An assessment of evidence from studies of five patient safety practices and proposals for research. *BMJ Quality and Safety*, 20, 604–610.

Øvretveit, J. and Staines, A. (2007). Sustained improvement? Findings from an independent case study of the Jönköping Quality Program. *Quality Management in Health Care*, 16(1), 68–83.

Parker, J. and Stanworth, H. (2005). 'Go for it!' Towards a critical realist approach to voluntary risk-taking. *Health, Risk & Society*, 7(4), 319–336.

Powell, A.E., Rushmer, R.K. and Davies, H.T.O. (2009). *A Systematic Narrative Review of Quality Improvement Models in Health Care*. Social Dimensions of Health Institute, The Universities of Dundee and St. Andrews.

Pronovost, P.J., Goeschel, C.A., Marsteller, J.A., Sexton, B.J., Pham, J.C. and Berenholtz, S.M. (2009). Framework for patient safety research and improvement. *Circulation*, 119, 330–337.

Rasmussen, J. (1997). Risk management in a dynamic society: A modelling problem. *Safety Science*, 27(2/3), 183–213.

Rasmussen, J. (2000). Human factors in a dynamic information society: Where are we heading? *Ergonomics*, 43(7), 869–879.

Rasmussen, J. and Svedung, I. (2000). *Proactive Risk Management in a Dynamic Society.* Karlstad, Sweden: Swedish Rescue Service Agency.

Righi, A.W., Saurin, T.A. and Wachs, P. (2015). A systematic literature review of resilience engineering: Research areas and a research agenda proposal. *Reliability Engineering & System Safety*, 141, 142–152. doi: 10.1016/j.ress.2015.03.007.

Ritzer, G. (1993). *The McDonaldization of Society.* New York: Sage Publications.

Sales, A. (2013). Quality improvement theories. In S. Straus, J. Tetroe and I.D. Graham (eds.), *Knowledge Translation in Health Care – Moving from Evidence to Practice*, 2nd edn. Wiley-Blackwell, pp. 320–328.

Schiff, G.D. and Rucker, T.D. (2001). Beyond structure–process–outcome: Donabedian's seven pillars and eleven buttresses of quality. *The Joint Commission Journal on Quality and Patient Safety*, 27(3), 169–176.

Shaqdan, K., Aran, S., Daftari Besheli, L. and Abujudeh, H. (2014). Root-cause analysis and health failure mode and effect analysis: Two leading techniques in health care quality assessment. *Journal of the American College of Radiology*, 11(6), 572–579.

Sherman, H., Castro, G., Hatlie, M. et al. (2009). Towards an international classification for patient safety: The conceptual framework. *International Journal for Quality in Health Care*, 21, 2–8. doi: http://dx.doi.org/10.1093/intqhc/mzn054.

Shewhart, W.A. (1931). *Economic Control of Quality of Manufactured Product.* New York: Van Nostrand.

Staines, A., Thor, J. and Robert, G. (2015). Sustaining improvement? The 20-year Jönköping quality improvement program revisited. *Quality Management in Health Care*, 24(1), 21–37. doi: 10.1097/QMH.0000000000000048.

Sutton, R. and Staw, B. (1995). What theory is not. *Administration Science Quarterly*, 40, 371–384.

Taylor, M., McNicholas, C., Nicolay, C., Darzi, A., Bell, D. and Reed, J. (2014). Systematic review of the application of the plan–do–study–act method to improve quality in healthcare. *BMJ Quality and Safety*, 23, 290–298.

Thor, J., Lundberg, J., Ask, J. et al. (2007). Application of statistical process control in healthcare improvement: Systematic review. *Quality and Safety in Health Care*, 16(5), 387–399.

UK Department of Health. (2000). *An Organization with a Memory: Report of an Expert Group on Learning from Adverse Events in the NHS.* London, U.K.: The Stationery Office.

Walshe, K. (2007). Understanding what works – and why – in quality improvement: The need for theory-driven evaluation. *International Journal for Quality in Health Care*, 19, 57–59. doi: 10.1093/intqhc/mzm004.

Wears, R.L., Hollnagel, E. and Braithwaite, J. (eds.). (2015). *Resilient Health Care*, Vol. 2: The Resilience of Everyday Clinical Work. Farnham, U.K.: Ashgate Publishing Ltd.

West, E. (2001). Management matters: The link between hospital organization and quality of patient care. *Quality in Health Care*, 10, 40–48. doi: 10.1136/qhc.10.1.40.

5 Working in an Institutionally Layered System on Patient Safety and Quality*

Hester van de Bovenkamp, Annemiek Stoopendaal and Roland Bal

CONTENTS

INTRODUCTION

In many Western healthcare systems, market elements, such as patient choice and transparency, have been introduced. Dutch healthcare is a case in point. Here, a system of regulated competition was officially introduced in 2006. The introduction of the regulated market did not mean that other previously existing institutional arrangements to govern healthcare quality, such as professional self-regulation and top-down state regulation, have disappeared. Instead, these pre-existing arrangements have become incorporated in and conditioned by regulated markets. This process of adding institutional arrangements to the old can be described as one of institutional layerings. Institutional layering in healthcare means that increasingly complicated mixed arrangements to govern healthcare quality are introduced (Streeck and Thelen 2005). Since the institutional context of healthcare organisations impacts quality and safety work and power relations in the healthcare sector, it is important to study this context. More specifically, when this context becomes more complex because

* This chapter is based on van de Bovenkamp, H., Stoopendaal, A., Bal, R. Working with layers: The governance and regulation of health care quality in an institutionally layered system, to be published and van de Bovenkamp, H., de Mul, M., Quartz, J., Weggelaar, A., Bal, R., Institutional Layering in governing health care quality, *Public Administration*, 92, 208–223, 2014.

of layering, of which the consequences can be unpredictable, it is important to learn more about how institutional layering works out in healthcare practice.

This chapter reveals the process of institutional layering in Dutch healthcare and shows the consequences this can have for actors working in such an institutionally layered context and how they in turn try to change this context. We specifically focus on the governance of quality of and in hospitals. The chapter is based on qualitative research into institutional layering of the governance of healthcare quality in the Netherlands and draws on interviews (n = 39) with key actors involved in the governance of healthcare quality, such as hospital directors, quality staff in hospitals, healthcare professionals, insurers, state regulators, policymakers and experts. In addition, we analysed relevant documents and literature on the governance of quality of care.

INSTITUTIONAL LAYERING

According to institutional theory, institutions shape the perception of problems and possible solutions. Moreover, institutional arrangements are important for setting incentives and constraints for action and shape power relations (March and Olson 1996; Mahoney and Thelen 2010). When analysing social processes, in this instance quality improvement (QI) work, it is therefore important to take the institutional context into account and try to understand this context in order to understand its influence on these processes.

To further our understanding of institutional arrangements, typologies have been developed (Helderman 2007; Pollitt and Bouckaert 2011). Table 5.1 is an example of such a typology. It builds on a distinction between four institutional arrangements (state, market, civil society and professional community) coupled with the governing mechanisms that are dominant within these arrangements (top-down regulation, contracts, consultation and self-regulation, respectively). The ideal types

TABLE 5.1
Ideal Typical Institutional Arrangements with Dominant Governing Mechanisms

		Level of Self-Regulation of Collective Actors	
		Low	**High**
Level of state intervention	**High**	State and hierarchy: top-down regulation	Civil society/association: consultation
	Low	Market: contract	Professional community: self-regulation

Sources: Based on Helderman, J.K., *Bringing the Market Back In? Institutional Complementarity and Hierarchy in Dutch Housing and Health Care*, Erasmus Universiteit Rotterdam, Rotterdam, the Netherlands, 2007; Bal, R., *De nieuwe zichtbaarheid: Sturing in tijden van marktwerking*, Erasmus Universiteit Rotterdam, Rotterdam, the Netherlands, 2008.

can be separated based on (1) the extent the state has a dominant position and is able to influence decision-making and (2) the room private and societal actors have to regulate themselves. Based on these axes, the typology separates four institutional arrangements which differ in terms of who is able to play a key role, how decision-making is organised and how decisions are implemented.

As said, the institutional arrangements in this table are ideal types. In practice, institutional arrangements are often hybrids; they consist of elements from more than one type. These ideal types therefore should be seen as a heuristic device that helps us to analyse the institutional arrangement and its changes (i.e. moving closer to or further away from a certain ideal type).

The concept of institutional layering has been introduced to better understand (the dynamic development of) complex hybrid arrangements (Streeck and Thelen 2005; Mahoney and Thelen 2010). Institutional layering occurs when new institutions are added to existing ones; new layers overlie others but do not necessarily replace them (Dent 2003; Pollitt and Bouckaert 2011). In other words, existing institutions still play a role, albeit a different one. Institutional layering can occur when challengers of the current system lack the capacity to change the original rules while defenders of the current system are unable to prevent new rules being added to the old (Streeck and Thelen 2005; Mahoney and Thelen 2010). Institutional layering does not mean that new arrangements simply overlie others but that arrangements interact with each other which can have important consequences for actors working in such a layered arrangement. For example, people working in a hospital on quality and safety will have to deal with the effects of working in a system that is both market driven and community based, asking them to compete *and* collaborate with other hospitals.

However, these actors are not just passively undergoing the effects of their institutional setting. They engage in institutional work which consists of the daily coping, keeping up and strategic use of their institutional context which can both sustain and change institutional structures (Feldman and Pentland 2003; Lawrence et al. 2011). In order to understand the consequences of layering for actors working in a layered healthcare system, it is therefore important to also look at the actions of these actors in response to their institutional context. In this chapter, we look at both the process of institutional layering in Dutch hospital care and how hospitals – more specifically, boards of directors which are responsible for the quality of care in hospitals – engage in institutional work.

INSTITUTIONAL LAYERING IN DUTCH HEALTHCARE

In this section, we describe the system of Dutch healthcare using the ideal types of institutional arrangements described earlier (Table 5.1).

In the Netherlands, as in most countries, quality of care was originally strongly based on *self-regulation by the professional community* (lower right quadrant Table 5.1). This arrangement drew mainly on education and on quality instruments such as peer review, clinical guidelines and visitation among peers (Klazinga 1996). The strong position of professionals to govern quality that is present in this arrangement proved not to be easily replaced. However, as we show in the following, throughout the years this position has been amended and increasingly other actors drawing on other instruments became involved in the governance of quality of care.

From the end of the 1980s onwards, *the civil society arrangement* in which *consultation with societal actors* involved in healthcare – professionals, healthcare providers, patient organisations, insurers and government – played an important role in order to get a grip on healthcare costs (top right quadrant Table 5.1) (Bal 2008). These consultations resulted in several acts to regulate quality, among which the Individual Healthcare Professionals Act (1997) and the Quality of Care Act (1996).

After several years, it was concluded that the implementation of these acts by healthcare organisations fell short (Casparie et al. 2001). For example, hospitals were meant to build quality systems which they did only to a limited extent. As a result of this, the question was raised if relying on self-regulation of the field was still adequate. This resulted on the one hand in more *state regulation* of quality by giving the Healthcare Inspectorate a more prominent role (top left quadrant Table 5.1). On the other hand, it resulted in the introduction of the *market-based system* in 2006 (lower left quadrant Table 5.1), which was from then on the official institutional arrangement to govern quality.* Within this system, providers, insurers and patients are meant to regulate healthcare through *competition* and *closing contracts* on three markets. In the healthcare provision market, hospitals (mostly private not-for-profit) compete for patients. In the healthcare insurance market (health insurance is obligatory), insurers compete for the insured. Both healthcare providers and insurers are active in the healthcare purchasing market. Here, insurers can buy care selectively on the basis of quality and price. Transparency of healthcare quality through the use of indicators is seen as a crucial instrument for the market to work, since quality information to make informed decisions is needed by patients, insurers and providers alike.

INTERACTIONS BETWEEN LAYERS

In this section, we describe how the introduction of the regulated market interacts with the other institutional layers the Dutch quality system is built on.

As stated earlier, *self-regulation by the professional community* has historically been strong. Respondents still recognise this as an important mechanism to ensure quality. However, they feel that relying on self-regulation is not enough. Some explicitly argue that outside pressure is needed for hospitals to work on QI, thereby legitimising other actors to play a role in governing quality.

> I have worked in hospitals for thirty years and it just does not happen [quality improvement from within]. (…) one way or the other it proves to be incredibly difficult. So therefore outside incentives are necessary to accomplish this or in any case to move substantially ahead and they [hospitals] will never succeed on their own.
>
> **Healthcare insurer**

* The introduction of the market arrangement was first and foremost an attempt to contain healthcare costs. However, the idea was that market parties would compete not only on price but also on quality, thereby giving healthcare providers an incentive to improve their quality.

The introduction of a system of regulated competition has caused the focus to shift from self-regulation (either by hospitals or by professionals) towards outside pressure for change. An important instrument for this is transparency of quality through the use of performance indicators. This transparency should allow patients to make choices about which hospital to go to, insurers to selectively buy care from hospitals and hospitals themselves to improve quality in order to stay ahead of the competition. The introduction of the market also means that other actors, such as insurers, patient organisations and private consultants, have increasingly become involved in QI. This outside interference does not mean that professional self-regulation is gone. It still plays a role through the quality instruments of the profession which are still in place (such as guidelines, peer review, medical education). The role these instruments play do start to change however as a result of the introduction of the market-based system since other actors also start to use them for regulatory purposes. Guidelines of the professional community, for example, were originally meant to guide choices of professionals who could decide, based on their professional autonomy, not to follow them. However, these guidelines are now also used by other actors such as the Inspectorate as standards for good quality care, which makes it harder for professionals to deviate from them when they think this is better for the patient. This way, instruments such as guidelines which originally were a way of safeguarding professional autonomy from outside regulation, can now, as they are used more stringently by other actors, actually partly erode this autonomy (e.g. Dent 2003).

The introduction of the regulated market also had its effect on the role of the *civil society* arrangement in which a *consensual mode of decision-making* is the dominant governance mechanism. Just as self-regulation, this institutional arrangement is not gone. However, it did become somewhat less important. Insurers, for example, argue that the consultation process takes too long to get results. Although they think that consultation is still important, they feel it is not absolutely necessary anymore and sometimes proceed with their QI efforts without consultation with the field. One respondent who works for an insurance company which published a ranking of hospitals with regard to breast cancer treatment explained that they chose to publish this list without consulting the field first. She explains the rationale behind this choice as follows:

> Yes changes have been made after that [the introduction of the breast cancer lists], and that mostly has to do with the involvement of professional and patient organizations. … The breast cancer trajectory has been developed very fast. On purpose, since we really wanted to make a statement: 'we are going to do this'. And you know as soon as you start to talk about it with someone, then everybody knows and then you do not get to make your point as effectively. So it was an explicit choice. However, now we feel we want to consult with professional and patient representative organizations beforehand; however, that does not mean that we will follow them no matter what.
>
> **Healthcare insurer**

The regulated market interacts with *top-down state regulation* in a different way. These institutional arrangements can best be described as mutually reinforcing since the introduction of regulated competition did not only put pressure on hospitals to improve but it also put pressure on government to become involved in the issue of healthcare quality. As a result of the increased transparency of healthcare quality, situations of failing quality have come to the fore (Klink 2009). These can often count on extensive attention from the media. In such a case, the Minister of Health is often called to account in Parliament. One response of the Ministry has been to stimulate markets parties to perform their designated role. An example of this is urging insurers to selectively buy their care based on quality (Klink 2009; Schippers 2011). A way to stimulate this process has been by increasing the number of diagnosis-related groups on which insurers can freely negotiate prices with hospitals (Schippers 2011). Another response to instances of failing quality coming to light has been to steer quality top down. One example of this is the Ministry's response to the publication of numbers on avoidable deaths:

> Then the minister said that his ambition was … that the number of avoidable deaths, whatever that is, would decrease by 50%. And there you see [the increased interference] of government, that was quite unique, government had not said something like that before … the minister had to explain to Parliament how something like that was possible.
>
> **Healthcare inspector**

A national safety programme forcing hospitals to have a safety system in place by the end of 2012 was initiated by the Ministry as the result of this debate (www.vmszorg.nl). Moreover, the Healthcare Inspectorate has increasingly taken on a more stringent role in the supervision of healthcare quality, closing hospital wards and surgeries which did not comply with the standards and putting an increasing number of organisations under 'stringent supervision'. Also, the Inspectorate has been ordered to use methods like mystery guests (in which case inspectors or members of the general public visit healthcare organisations pretending to be family members of future clients in order to get a view of day-to-day care provision) and unannounced visits (in which case inspectors visit healthcare organisations unannounced) in order to get more grip on healthcare quality (Schippers 2012; Adams et al. 2015; Stoopendaal 2015). In addition, the Ministry has set up a national quality institute, which is meant to stimulate guideline development, in order to gain more control over quality (Klink 2010; https://www.zorginstituutnederland.nl/kwaliteit).

We can conclude that, as the term 'regulated competition' already seems to imply, state and market arrangements are mutually reinforcing. Nevertheless, to say that state and market are now the dominant arrangements is overstated, as self-regulation and consultation are still very much at play as well, albeit with different meanings attached. This means that the institutionally layered system that is in place at the moment has caused many actors to be involved with the issue of quality of care in hospitals, who all use their own instruments to govern quality or each other's instruments in different ways. A summary of the layered institutional arrangement can be found in Table 5.2.

TABLE 5.2
Layered Institutional Arrangement Dutch Healthcare

Institutional Arrangement	Important Actors	Steering Instruments	Period
Market	Insurers, healthcare providers, patients	Competition, closing contracts, transparency	Officially introduced as the dominant arrangement in 2006, after an incremental change process
State and hierarchy	Ministry of Health, Healthcare Inspectorate	Top-down regulation through legislation (e.g. Quality of Care Act, Individual Healthcare Professions Act) and supervision	Always played a role, importance increased from the 2000s onwards and especially after the implementation of the market-based system
Civil society/ association	Healthcare professionals, providers, insurers, patient organisations, government	Consultation and deliberation, for example, in setting performance indicators. Was important in the development of the Quality of Care Act in the 1990s. And again in setting limits on economic growth of the healthcare sector in 2012	Especially important in the 1990s, still plays a role but less dominant than before
Professional community	Healthcare professionals	Medical training, peer review, guidelines, visitation, quality systems	Oldest, still highly important but less dominant than before

WORKING WITH LAYERS

The layered arrangement means that hospitals have to deal with many actors with diverging quality demands ranging from top-down government legislation, supervision and Q&S programmes; to negotiations with insurers; to visitation, guidelines and education of professional organisations and to patient organisations that have developed their own indicators, prizes and rankings. Moreover, due to the increased transparency of quality of care, the media also plays an increasing role in the quality debate, as do politicians who are responding to whatever is reported in the media. Hospitals have to relate to all these actors and their quality and information demands within differing governance regimes (market negotiations, state control, consultation and self-governance). In the following, we discuss how hospital boards engage with the layered institutional context and how they try to use and influence this context.

Our respondents emphasise the important role transparency, introduced as part of the market-based system, and the resulting media attention for quality incidents played in getting the subject of quality on the agenda of hospital boards.

> The responsibility for the content of care was completely assigned to professionals; they did not have to be accountable for this. This was changed legally with the introduction of the Quality of Care Act in 1996. The Boards of Directors legally carry the final

> responsibility for everything in the organization. That includes the quality and safety of the care process. That has been duly noted for a long time. Nothing happened with that however. Until, and that is the first time this was exposed, the Radboud case [case concerning high mortality rates on a cardiology ward that lead to public upheaval in 2008]
>
> **Governance expert and former chief inspector**

Incidents like these were analysed in terms of failing governance by both the Inspectorate and the Ministry of Health. For example, the Ministry of Health emphasised the role and responsibility of the Supervisory Boards and the 'indisputable end responsibility' of the Board of Directors in response to cases of failing quality (Klink 2009; Schippers and Van Rijn 2013), reframing patient safety as a managerial and governance issue (Behr et al. 2015). So even though the Board of Directors was assigned the legal responsibility for ensuring quality of care already in the mid-1990s, this was put into practice much later, when as a result of the introduction of the *regulated competition system* accompanied by the increased transparency of care, they were urged to take up their role. This role attributed to hospital boards in ensuring quality of care has consequences for the internal power relations in hospitals. The position of hospital boards vis-à-vis professionals has been strengthened as a result, a relationship which is described as a continuous struggle.

> If the relationship [between Board of Directors and medical professionals] is poor, then the medical staff will say to the director: quality in the consultation room is our thing. [I] agree when this concerns the one-on-one contact with the patient. [I] disagree when this concerns what happens next. I don't need to know what happens in the consultation room, but I do need to know how the individual specialist and his group of specialists work on quality and how they ensure the quality of care provided in that consultation room
>
> **Former hospital director and former inspector**

Boards of Directors draw on the institutionally layered context to get more grip on quality and by consequence on healthcare professionals. In order to gain control, they, for example, built quality systems for which they draw on the myriad of information that is available as a result of the layered arrangement with its information demands of external actors. They also use instruments used by other actors to strengthen their position internally. For example, Bal et al. (2015) have shown that hospital rankings, which are part of the market-based system, are used internally by hospital boards as a strategic tool to weaken the position of medical professionals and to take up their responsibility with regard to quality. The information drawn from indicators is used in a similar way.

> Throughout the years you step in earlier (…). That helps. Because there is more information, we now at least have the HSMR [Hospital Standardized Mortality Rate], well we did not have those eight years ago.
>
> **Hospital director**

The layered institutional arrangement does not only help hospital boards to take up their responsibility with regard to quality but also poses new problems for them.

Being in an institutionally layered arrangement means having to relate to many external actors who all want, slightly different, information. First, they have to relate to market parties such as patient organisations and insurers. Hospital boards do not always see the added value for quality of this.

> They poor out this bureaucracy all over you and you have to comply with that. So that keeps your organization busy. And subsequently the health care purchaser [insurer] comes by and you just talk about money.
>
> **Hospital director**

Second, they have to relate to quality demands of the professional associations that are active in the self-regulatory regime. Third, they have to relate to state legislation and the Inspectorate who, as shown earlier, are also increasingly involved in quality issues. Especially because government takes action in an incident-driven way this causes problems according to respondents since this formalisation and fragmentation prevents them to work on quality in a structured manner. Moreover, the fact that these external actors rely heavily on formal systems in which hospitals have to account for their quality work, makes it hard for hospital boards to incorporate more informal quality assurance activities, such as visiting wards and having informal conversations with staff. These informal activities are, according to respondents working in hospitals, highly important to govern quality. However, they are hard to give account of to external parties. This causes the formalistic tools used by external parties and the need for situated informal approaches to be increasingly disconnected.

In response to these challenges, hospital boards perform institutional work in order to try to influence the institutional context they are in. One strategy they use is to refuse to comply with external demands when boards feel they are of no extra value to quality of care. They argue that they have to prioritise certain demands over others to keep the pressure manageable and try to regain control over the quality agenda (see also Quartz et al. 2012).

> They [a hospital] made an inventory about the number of guidelines, regulation etc. they had to follow and they came to the number of 1400. His [director of the hospital] proposition is that you have to make a well argued choice about what you focus your attention on, because 1400 subjects is a pointless exercise (…) that you pick ten themes and as Board of Directors you decide to work on those.
>
> **Healthcare Inspectorate**

Hospital boards also try to deal with quality more proactively in an attempt to reshape and (re)gain control over quality. They do so, for instance, by building quality systems which pay attention to both formal and informal aspects of quality governance, in this way overcoming the overly formalistic approach used by outside actors. Last, by actively contributing to national discussions on quality and development processes of quality instruments, hospital boards and professionals also try to influence the external demands put on them through the mechanisms of the associational order. These examples show that institutional work can help actors to influence their position in relation to the institutional context.

CONCLUSIONS: LESSONS FROM THE DUTCH CASE

The institutional context of healthcare organisations impacts their quality and safety work and the power relations of the organisation vis-à-vis external parties as well as of hospital boards vis-à-vis professionals in their organisations. When analysing quality and safety work, it is therefore important to pay attention to this institutional context.

In healthcare, these institutional contexts have become increasingly complex due to the twin introduction of market mechanisms, such as patient choice and transparency and an increasing influence of the state, which have complemented professional self-regulation. This increased complexity can also be seen in the Nordic countries (Deilkås et al. 2015; Lehman et al. 2015; Øvretveit et al. 2015). In this chapter, we focused on the complexity of these governance arrangements, its consequences and how hospitals relate to these by using the concepts of institutional layering and institutional work. We used Dutch healthcare as a case study. The specifics of the layered arrangements will be different across countries. For instance, in the Nordic countries historically a consensual mode of decision-making has been more dominant than in other countries, which will impact the implementation of new policies (van de Bovenkamp et al. 2011). However, this does not mean that we cannot draw lessons from this chapter for other healthcare systems.

First, an important lesson is that how reforms such as the introduction of market mechanisms in healthcare work out will depend on their interaction with existing institutional arrangements (see also Dent 2003; Kodate 2010; Pollit and Boackaert 2011). The effects of these reforms partly depend on the local dynamics created by the history of the healthcare system. We therefore need comparative studies between countries to understand these dynamics and effects better. This insight also provides a lesson for policymakers and analysts who need to think about and reflect on the interactions between arrangements and the effects certain reforms create.

Second, we can learn that actors working in hospitals (or other healthcare organisations affected by layering) have to deal with the effects of institutional layering such as fragmentation of quality demands. Our case study offers important insights into how they can do this, namely, by using the new institutional arrangements to change power relations (e.g. between hospital boards and professionals) or to influence new institutional frameworks by using pre-existing ones (such as we saw in the attempts of hospitals to influence and negotiate external demands).

The chapter shows that having to relate to an institutionally layered context can be both enabling and disabling for actors working on quality in healthcare organisations, in our case hospital boards. On the one hand, it has given them the opportunity to strengthen their position on quality internally and also given them the instruments to better govern quality within the hospital organisation. Hospital boards can therefore also use the layered context to further their end. On the other hand, it has provided them with additional challenges to work on quality effectively. It has been shown before that layering can have a burdening effect (Van der Heijden 2011). Our case supports this conclusion since having to relate to external actors involved in healthcare quality can result in working on quality in an ad hoc manner because of a politics of incidents, a high registration burden and a strong focus on accounting for one's actions which draws attention away from the informal activities that are needed for actors to take up their role. We have also seen, however, that actors engage in institutional work to reshape their institutional context in order to work on quality more effectively (Lawrence et al. 2011). Working in a layered institutional arrangement is therefore not only a struggle for actors, but it also offers potential to use and play with layers in order to get a grip on healthcare quality and safety.

In order to use this potential, insight into the specifics of layering of a certain system is highly relevant. The same goes for the specifics of how actors can try to influence their institutional context, which may differ between countries. This insight is highly relevant since healthcare organisations and policymakers alike can use it to govern quality more effectively. 'Archaeological' studies into layering and studies into the institutional work of actors can help in gaining such insights.

REFERENCES

Adams S, Paul K, Ketelaars C, Robben P. 2015. The use of mystery guests by the Dutch Health Inspectorate: Results of a pilot study in long-term intramural elderly care. *Health Policy*, 2(119): 125.

Bal R. 2008. *De nieuwe zichtbaarheid: Sturing in tijden van marktwerking.* Rotterdam, the Netherlands: Erasmus Universiteit Rotterdam.

Bal R, Quartz J, and Wallenburg I. 2015. Making hospitals governable. How rankings are transforming hospital organizations. IRSPM, Bimingham, U.K.

Behr L, Grit K, Robben P, Bal R. 2015. Framing and re-framing critical incidents in hospitals. *Health, Risk & Society*, 17, 81–97. doi: http://dx.doi.org/10.1080/13698575.2015.1006587.

Casparie AF, Legemaate J, Rijkschroeff RAL, Brugman MJE, Buijsen MAJM, Hulst EH et al. 2001. *Evaluatie Kwaliteitswet zorginstellingen.* Den Haag, the Netherlands: ZonMw.

Deilkås T, Ingebrigststen T, Ringand A. 2015. Norway, in: *Healthcare Reform, Quality and Safety: Perspectives, Participants, Partnerships and Prospects in 30 Countries,* Braithwaite J, Matsuyama Y, Mannion R, Johnson J (eds.), pp. 261–271. Farnham, U.K.: Ashgate.

Dent M. 2003. *Remodelling Hospitals and Health Professions in Europe: Medicine Nursing and the State.* New York: Palgrave Macmillan.

Feldman MS, Pentland BT. 2003. Reconceptualizing organizational routine as a source of flexibility and change. *Administrative Science Quarterly*, 48, 94–118.

Helderman JK. 2007. *Bringing the Market Back In? Institutional Complementarity and Hierarchy in Dutch Housing and Health Care.* Rotterdam, the Netherlands: Erasmus Universiteit Rotterdam.

Klazinga NS. 1996. *Quality Management of Medical Specialist Care in the Netherlands: An Explorative Study of Its Nature and Development.* Rotterdam, the Netherlands: Erasmus University Rotterdam.

Klink A. 2009. *Ruimte en rekenschap voor zorg en ondersteuning.* Den Haag, the Netherlands: Ministerie van Volksgezondheid Welzijn en Sport.

Klink A. 2010. *Oprichting van een nationaal kwaliteitsinstituut.* Den Haag, the Netherlands: Ministerie van Volksgezondheid Welzijn en Sport.

Kodate N. 2010. Events, public discourses and responsive government: Quality assurance in health care in England, Sweden and Japan. *Journal of Public Policy*, 30, 263–289.

Lawrence T, Suddaby R, Leca B. 2011. Institutional work: Refocusing institutional studies of organization. *Journal of Management Inquiry*, 20, 52–58.

Lehman KJ, Engel C, Eriksen J. 2015. Denmark, in: *Healthcare Reform, Quality and Safety: Perspectives, Participants, Partnerships and Prospects in 30 Countries*, Braithwaite J, Matsuyama Y, Mannion R, Johnson J (eds.), pp. 215–226. Farnham, U.K.: Ashgate.

Mahoney J, Thelen K. 2010. *A Theory of Gradual Institutional Change*. Cambridge, U.K.: Cambridge University Press.

March JG, Olsen JP. 1996. Institutional perspectives on political institutions, *Governance: An International Journal of Policy and Administration*, 9(3), 247–264.

Øvretveit J, Sachs MA, Lindh M. 2015. Sweden, in: *Healthcare Reform, Quality and Safety: Perspectives, Participants, Partnerships and Prospects in 30 Countries*, Braithwaite J, Matsuyama Y, Mannion R, Johnson J (eds.), pp. 285–296. Farnham, U.K.: Ashgate.

Pollitt C, Bouckaert G. 2011. *Public Management Reform: A Comparative Analysis, New Public Management, Governance and the Neo-Weberian State*. Oxford, U.K.: Oxford University Press.

Quartz JGU, Weggelaar AM, van de Bovenkamp HM, Bal R. 2012. *Quality & Safety in Europe by Research (QUASER) Country Report: The Netherlands*. Rotterdam, the Netherlands: iBMG/Quaser.

Schippers E. 2011. *Zorg die werkt: de beleidsdoelstellingen van de minister van Volksgezondheid, Welzijn en Sport*. Den Haag, the Netherlands: Ministerie van Volksgezondheid Welzijn en Sport.

Schippers E. 2012. *Toezicht op de gezondheidszorg door de IGZ*. Den Haag, the Netherlands: Ministerie van Volksgezondheid Welzijn en Sport.

Schippers E, Van Rijn M. 2013. *Gezamenlijke agenda VWS: 'Van systemen naar mensen'*. Den Haag, the Netherlands: Ministerie van Volkgsgezondheid Welzijn en Sport.

Stoopendaal A. 2015. Mystery Guests 2: Begeleidend evaluatie onderzoek vervolgproject IGZ-Ouderenzorg, instituut Beleid & Management Gezondheidszorg, Rotterdam, the Netherlands.

Streeck W, Thelen K. 2005. *Beyond Continuity: Institutional Change in Advanced Political Economies*. Oxford, U.K.: Oxford University Press.

van de Bovenkamp H, Stoopendaal A, Bal R. Working with layers: The governance and regulation of health care quality in an institutionally layered system, to be published.

van de Bovenkamp HM, de Mul M, Quartz J, Weggelaar-Jansen AM, Bal R. 2014. Institutional layering in governing healthcare quality. *Public Administration*, 92, 208–233.

van de Bovenkamp HM, Quartz J, Weggelaar-Jansen AM, Bal R, on behalf of the QUASER team. 2011. Guiding quality work in European hospitals, iBMG: Rotterdam, the Netherlands.

van der Heijden J. 2011. Institutional layering: A review of the use of the concept. *Politics*, 31, 9–18.

www.zorginstituutnederland.nl/kwaliteit. Accessed 11 April, 2016.

www.vmszorg.nl. Accessed 11 April, 2016.

Section II

Contemporary Nordic Research – Macro-Level Issues

6 Centralisation Efforts to Improve the Quality of Care and Reduce the Costs in Healthcare Systems

Pia Kjær Kristensen and Søren Paaske Johnsen

CONTENTS

INTRODUCTION

Efforts to improve the quality of healthcare and reduce costs in healthcare systems often focus on centralising care in high-volume hospital units. Worldwide hospital consolidation has increased over the last few years; in the United States, it is predicted that 20% of all hospitals will seek a merger in the next 5 years (Creswell and Abelson 2013). In some Nordic countries (e.g. Denmark), some level of centralisation has already occurred; but in other countries, this process has been less pronounced to date, due to challenges like isolating geography or long distances between hospitals (e.g. Norway).

Several studies have investigated whether the number of patients treated in a department (or the number of specific procedures performed) was associated with patient outcome. This association, called the volume–outcome relationship, has received increasing interest over the decades, due to the hypothesis that 'practice makes perfect' and an expectation that centralised care may lead to more cost-effective care, due to the economics of scale. Indeed, large enterprises generally gain cost advantages, because fixed costs can be spread out over more units of output; therefore, the cost per unit of output decreases with increasing volume.

This chapter describes the current literature on the volume–outcome relationship, including a discussion on the methodological issues related to interpreting reported data. Furthermore, we will describe potential mechanisms underlying this association, including the quality of delivered care, clinical skills, patient selection and hospital organisation. Finally, we will discuss interpretations of the data available on the volume–outcome relationship and the implications of healthcare centralisation, particularly for Nordic countries.

SCALE EFFECTS

A large number of studies have investigated the volume–outcome relationship across a wide range of clinical conditions. This relationship has been most frequently studied with regard to surgery and cancer care, but some studies investigated this relationship with regard to patients with acute and chronic medical conditions. In the following review, we will analyse the current understanding of the volume–outcome relationship across different diseases to assess potential scale effects. The scale effect most often pertains to cost-effectiveness, but it may also pertain to quality of care improvements, due to increased routine and specialisation in the organisation.

Definition of Volume

Volume can be defined both as the cumulative number of patients ever treated (e.g. as in a learning curve for a surgeon learning a new technique) and as the number of patients that must be treated per time period to maintain proficiency in a skill or procedure. Later, we will focus on the latter definition. Moreover, the literature distinguishes between the volume pertaining to a single physician and the volume pertaining to a hospital. Thus, the 'physician volume' is the experience required for an individual physician to provide satisfactory quality of care with a given technique or procedure (e.g. a surgeon that performs a few procedures per year is likely perform surgery less effectively than a surgeon that performs several procedures per week). In contrast, the 'hospital volume' reflects the hospital's overall experience with a procedure (e.g. high-volume units may be more likely to have teams that work effectively together, systems in place to identify complications early and efficient mechanisms for responding to complications).

It is likely that the provider (e.g. physician) volume and hospital volume interact, but few studies have examined both types of volume simultaneously. One of those studies found a multiplicative effect; thus, better outcomes were observed when high-volume physicians practised in high-volume hospitals (Hannan et al. 1991). However, other studies have not confirmed that effect (Lindenauer et al. 2006).

Very few studies have investigated patient volume as a continuous variable. In contrast, most studies had used fixed categories of patient/procedure volume. Due to the different organisations found among healthcare systems internationally, a variety of volume categories have been used. Consequently, volume that may be defined as high in one study may correspond to a low volume in another study. Therefore, caution is necessary when making comparisons across studies.

Outcomes

Mortality is the most frequently examined outcome. Few studies have examined the risk of complications, length of hospital stay or risk of readmission (Halm et al. 2002). However, the possible association of volume with these outcomes is important, because they represent aspects of healthcare other than mortality. For many conditions, such as patients undergoing knee and hip arthroplasties, mortality is not a useful outcome measure, due to the very low mortality risk (Glassou et al. 2014). Complications, such as the revision rate, might be a more appropriate measure, because it is a frequent outcome after hip/knee replacement surgery, and it may more directly reflect the quality of provided care. However, for many studies, the follow-up period is too short to detect all relevant complications. In addition, the length of hospital stay and readmission rate reflect both in-hospital care and care transitions between hospitals and primary care.

Scale Effects across Different Diseases

Many studies have linked a high patient volume with low mortality, particularly for patients that require advanced surgical and invasive procedures (Halm et al. 2002, Koelemay and Vahl 2007, Post et al. 2010, Ross et al. 2010, Joynt et al. 2011, Zevin et al. 2012). For example, patients with cancer of the oesophagus, liver and pancreas appear to have better survival rates when they are admitted to high-volume units or treated by high-volume physicians than when they are treated at low-volume units or by low-volume physicians (Begg et al. 1998, Gordon et al. 1999). The same association was observed in patients that underwent treatments for cardiovascular diseases, such as percutaneous coronary intervention, surgery for abdominal aortic aneurysms, thrombolytic therapy for myocardial infarction, paediatric cardiac surgery or bypass surgery (Halm et al. 2002, Koelemay and Vahl 2007, Post et al. 2010, Joynt et al. 2011). Patients that receive organ transplantation or bariatric surgery at high-volume hospitals also appear to have better survival rates than patients admitted to low-volume hospitals (Zevin et al. 2012). However, evidence that supports a relationship between volume and outcome is quite limited and inconsistent for many relatively simple surgical procedures and conditions, for example, orthopaedic procedures (Halm et al. 2002, Franzo et al. 2005, Kristensen et al. 2014). Although surgical procedures have received more attention than medical treatments, there is some evidence to suggest that patients with acute myocardial infarction, breast cancer, stroke or HIV infections benefit from physicians that have treated more patients with those conditions (Bennett et al. 1989, Sainsbury et al. 1995, Ross et al. 2010, Svendsen et al. 2012). One study that focused on common medical conditions, including pneumonia, showed no difference in mortality between low-volume and high-volume hospitals (Lindenauer et al. 2006).

Overall, existing studies have indicated that a high volume may be associated with low mortality among patients that undergo advanced surgery or invasive procedures and among patients with complex medical conditions. In contrast, no consistent association has been found for patients that undergo relatively simple, standardised surgical procedures.

The association is less clear between patient volume and other outcomes, including length of stay and readmission rates. Among patients with hip fractures and patients with pneumonia, a high volume has been associated with a longer length of stay (Lindenauer et al. 2006, Kristensen et al. 2014), but in other patient groups, such as stroke, high volume has been associated with a shorter length of stay (Svendsen et al. 2012). This variation in findings is striking. One potential interpretation is that patients undergoing advanced invasive procedures or patients with complex medical conditions benefit most from high-volume facilities; however, the results may have been skewed by unmeasured factors, such as differences in discharge policies. Alternatively, an association between a high volume and a short length of stay could also arise from the reverse relationship, because a shorter length of stay would result in more beds available and thus a larger volume of patients.

A few studies have examined readmission as an outcome. Of these, some found that among patients undergoing major surgery, increased volume was associated with reduced readmission rates (Nguyen et al. 2004, Tsai et al. 2013); however, others could not confirm this relationship (Goodney et al. 2003, Borenstein et al. 2005). To our knowledge, only one study has examined the association between volume and readmission among medical patients. In that study, hospitals with low volumes had lower standardised readmission rates than hospitals with high volumes (Horwitz et al. 2015).

In conclusion, there does not seem to be a general association between volume and outcome across diseases, even when considering different outcomes, including mortality, length of stay and readmission. Nevertheless, there is an indication that patients undergoing advanced invasive procedures or patients with complex medical conditions may benefit most from a facility with a high patient volume.

Existing studies have had a number of methodological limitations, including the risk of bias, inadequate control for mixed cases, lack of hierarchical modelling to account for clustering (not taking into account that patients within the same hospital and unit are correlated) and moderate statistical precision. All studies have used an observational design; therefore, it is important to consider the selection procedures and risk adjustments for differences in illness severity and co-morbidity, also called the case mix. Although case-mix adjustments may improve the likelihood of fair comparisons between hospitals, there may be unknown biases that are impossible to adjust for; therefore, the results should be interpreted with caution. Lacking hierarchical modelling in the statistical analysis to account for clustering can produce spuriously significant effects, because clustering affects the sampling variance. Moreover, statistical precision can be a problem. Patient samples may be too small to achieve statistically significant differences, or even when significant, the volume–outcome relationship may not show sufficient strength for clinical relevance (i.e. large sample sizes may produce results that are statistically significant, but the absolute differences may only be minor).

IS THERE A LINEAR SCALE ADVANTAGE?

In addition to the substantial heterogeneity in findings between studies, variation is also observed in findings within individual studies. For example, some patients admitted to low-volume hospitals may experience excellent outcomes, and patients admitted to high-volume hospitals may experience poor outcomes. Furthermore, the effect of volume may be reduced over time, as procedures become well established. These inconsistencies could indicate that the volume of patients is not a mechanism in itself, but rather a contextual descriptor, which is not directly related to outcome (Jha 2015). Several potential mechanisms may underlie a volume–outcome relationship. In a report from the Institute of Medicine, entitled 'Interpreting the Volume-Outcome Relationship in the Context of Health Care Quality' (Hewitt 2000), a conceptual model was drafted to explain the contextual mechanisms that can influence the relationship between volume and outcome. A modified version of that model is presented in Figure 6.1.

The model illustrates important contextual mechanisms that influence the volume–outcome relationship. Outcome is associated with patient selection, including disease severity and co-morbidity, the processes of care and the skills that pertain to both hospitals and clinicians; all of these factors may be affected by volume. In the following sections, we will describe the elements in the model and their relationships to volume.

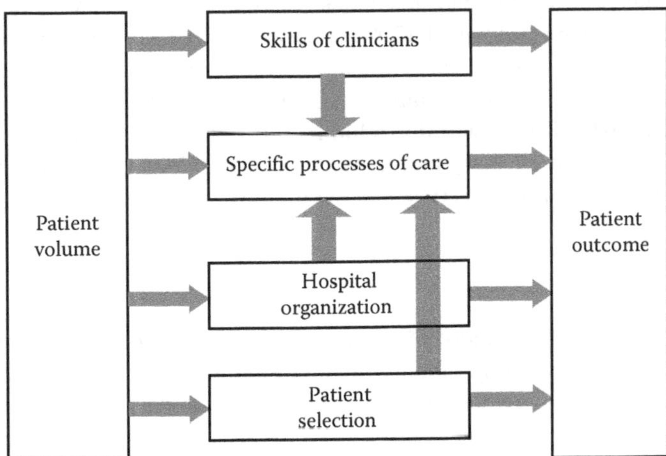

FIGURE 6.1 Contextual mechanisms that influence the volume–outcome relationship. (Based on Hewitt, M.E., *Interpreting the Volume-Outcome Relationship in the Context of Health Care Quality: Workshop Summary*, Institute of Medicine, Washington, DC, 2000.)

Patient Selection

Different types of patients may be admitted to low-volume and high-volume hospitals. A study by Liu et al. (2006) found substantial differences in the characteristics of patients admitted to hospitals with different volumes. They showed that, in general, blacks, Asians, Hispanics, patients that receive Medicaid (a federal programme in the United States that provides assistance with medical costs to people with limited income and resources) and uninsured patients were less likely to go to high-volume hospitals. Because these patients may have higher levels of disease severity and co-morbidities, patients admitted to low-volume hospitals may have a worse prognostic profile than patients admitted to high-volume hospitals (Liu et al. 2006). In contrast, another study reported that high-volume surgeons were more likely to perform inappropriate carotid endarterectomies than low-volume surgeons; that finding indicated that patients at high-volume units may have a higher risk profile than patients at low-volume units (Brook et al. 1990). In some cases, patients with a high-risk profile may be referred to high-volume units for specialist treatment; however, in other cases, patients with a high-risk profile might be transferred to low-volume units for end-of-life care (e.g. after specialised treatments have failed). Another type of patient selection that can influence the volume–outcome relationship is known as selective referral. In this case, more referrals are directed to excellent physicians and hospitals with a good reputation for achieving better outcomes; thus, a high volume may be associated with better outcomes.

Varying attention has been directed towards adjusting for disease severity and co-morbid conditions. The methods used for these adjustments may affect patient outcome evaluations. In addition, the quality of patient information may affect whether an adjustment for case mix is performed. For example, high-volume hospitals are more research intensive, and therefore, they may have collected more patient information on a routine basis. Thus, high-volume hospitals may have greater capabilities in adjusting for patient-related risk factors. This notion was supported by findings in a review by Halm et al. (2002). They stated that studies that included risk adjustments based on clinical data were less likely to report significant associations between patient volume and mortality than studies that adjusted for risk based on administrative data. They noted that administrative data validity could vary across different patient volumes; thus, administrative data provided better case-mix adjustments at high-volume hospitals than at low-volume hospitals (Halm et al. 2002).

Quality of Care

Most studies lack an examination of the mechanisms and specific processes involved in how volume might directly influence outcome. However, the association between volume and quality of care has been examined in a limited number of studies.

In studies where high-volume units were associated with better outcomes, the association appeared to be mediated through high-quality care. In a study on patients with acute myocardial infarctions, the differential use of treatments recommended by guidelines, including treatments that were proven effective (thrombolytics, aspirin, beta blockers and angiotensin-converting enzyme inhibitors), could explain about one-third of the survival benefit attributable to high-volume hospitals (Thiemann et al. 1999). They showed that patients at high-volume hospitals were more likely to receive treatments recommended by guidelines, including effective pharmaceutical treatments, compared to patients treated at low-volume hospitals. Another study on patients that underwent rectal cancer surgery found that high hospital volumes were associated with improved global quality scores and low 30-day mortality (Leonard et al. 2014).

In other studies, where high-volume units were associated with poor outcome, the association appeared to be mediated by low-quality care. In a study of patients with hip fractures, a higher patient volume was associated with higher 30-day mortality, longer length of stay and a lower quality of in-hospital care (Kristensen et al. 2014). In addition, admission to a low-volume unit was associated with a higher probability of receiving the treatment recommended by guidelines. Thus, the different effects of volume on 30-day mortality appeared to be partly explained by differences in the quality of in-hospital care, due to greater adherence to guidelines (Kristensen et al. 2014). This association between patient volume and providing the care recommended by guidelines was also observed in a study of patients with pneumonia. Lindenauer et al. (2006) found that, as unit volume increased, the risk of delay in antibiotics administration also increased. Consistent with that finding, another study in emergency departments found that increases in the total number of emergency department visits were associated with decreases in adherence to guideline-recommended diagnostic tests; thus, higher numbers were associated with a higher risk of exacerbation among patients with pulmonary disease (Tsai et al. 2009).

Quality of care, reflected by whether patients receive the recommended care processes, seems to be an important intermediate factor for the volume–outcome association. However, not all studies have found this association.

A study on patients that received coronary artery bypass surgeries found that controlling for quality of care differences did not have any impact on the association between volume measures and mortality. However, adherence to process quality measures was associated with improved mortality rates, independent of hospital or physician volume (Auerbach et al. 2009). In another study on patients with strokes, a higher patient volume was associated with improved quality of care and reduced hospital bed day use, but it was not associated with mortality. Adjusting for differences in care quality had no major influence on mortality or on the bed day use (Svendsen et al. 2012).

These studies indicated that other aspects of care or other mechanisms that were not investigated may have contributed to the outcomes obtained. Several underlying mechanisms might explain the volume–outcome relationships, including clinical skills and hospital organisation.

Clinical Skills

Different levels of skills possessed by treating physicians may also be associated with patient outcomes. For example, in surgical procedures, surgeon dexterity is often a critical factor in minimising trauma to adjacent tissues; that is, an inexperienced surgeon may cause more surgical trauma than an experienced surgeon (Palm et al. 2007, Liu et al. 2015). Furthermore, cognitive skills may be important in producing high-quality outcomes, such as the ability to recognise uncommon signs of medical conditions and take appropriate, timely action. In addition, we could hypothesise that adherence to current guidelines is also related to clinical skills (Hwang et al. 2013). For instance, team members that are familiar with the disease treatment guidelines may be predisposed to recommending the appropriate treatment, because they utilise the guidelines on a regular basis.

Hospital Organisation

A third possible mechanism that can complicate the interpretation of the volume–outcome relationship includes system-level factors, such as hospital organisation. Organisational skills, including the ability of the system to facilitate effective care (e.g. multidisciplinary coordination and the ability to respond effectively to complications), may contribute to patient outcome (Bates and Gawande 2003). Furthermore, high-volume hospitals may be more likely to have state-of-the-art equipment that can contribute to avoiding errors (e.g. sophisticated computerised reminder systems and advanced diagnostic equipment). In addition, high-volume hospitals may have the financial capacity to employ teams specialised in home-based follow-up or patient education programmes, which may improve outcome (Swart et al. 2015). On the other hand, the high activity levels present in the large, complex environments of high-volume hospitals may pose a challenge to implementing and perfecting fundamental care.

INTERPRETING THE VOLUME–OUTCOME RELATIONSHIP

The absence of a consistent positive relationship between high volume and quality of care suggests that the benefits associated with greater disease experience in busy hospitals could be outweighed by the challenges of implementing and perfecting care in large, complex environments. Furthermore, we cannot exclude the existence of publication bias, which would diminish the likelihood of identifying studies that failed to find a volume–outcome association.

Volume Threshold

An interesting question is whether there might be a volume threshold for the volume–outcome relationship. In other words, there may be a volume range that shows the effect, but once a certain volume is reached, increasing the volume would no longer be associated with better outcomes.

Indeed, for medical conditions, one study identified an annual volume threshold, above which, increases in annual volume no longer significantly affected outcome (Ross et al. 2010). That study showed that the curves representing the association between volume and risk-adjusted mortality began to level out at different thresholds for different conditions. The annual volume thresholds were 610 patients for myocardial infarction, 500 patients for heart failure and 210 patients for pneumonia.

Those results suggested that the benefit of an increased patient volume would be most pronounced at hospitals with fewer than 100 cases of a given condition. The same pattern was observed in studies on how the volume of patients with hip fractures was related to 30-day mortality in countries where some centralisation had already occurred (Franzo et al. 2005, Kristensen et al. 2014). In those studies, a high patient volume was defined as greater than 352 patients with hip fractures per year. They showed that high volumes were associated with higher 30-day mortality. In other patient groups, such as patients with stroke, a graph suggested that the advantages of high volume were most evident in stroke units that treated up to 300–400 patients annually (Svendsen et al. 2012).

The finding of a volume threshold may arise from the fact that the volume effect may be diminished or even reversed in large-scale institutions. Large-scale institutions may have several disadvantages that counteract the benefits of volume, including deficient coordination, increased bureaucracy and an unrestricted access to expensive but not necessary relevant technology, due to specialisation (Posnett 1999, Kristensen et al. 2010).

IMPLICATIONS

The volume–outcome relationship has been one of the key arguments for centralisation of healthcare, despite the weak evidence to support an association for many medical conditions. Most volume–outcome studies have been performed on data collected prior to 2000, when hospital consolidation had only occurred to a limited extent in many settings. Therefore, it is uncertain whether we should generalise those study findings to the current healthcare systems, which may have substantially higher levels of centralisation. In particular, there appears to be a lack of evidence to support a positive association between increasing patient volume and improved outcome among patients with common medical conditions that require low-risk invasive procedures. The literature has also indicated that volume thresholds may exist; thus, increasing the volume may not be associated with improved outcomes or may even lead to worse outcomes.

In addition, there is limited evidence regarding the relationship between very high volumes and cost. The association between volume and cost has been investigated, but as mentioned earlier, it is not known whether a potential threshold exists (Bardach et al. 2004, Slattery et al. 2004, Losina et al. 2009, Tsugawa et al. 2013, Albornoz et al. 2014). One potential hypothesis might be that some organisational elements, such as surgical teams and emergency department teams, would be better utilised in high-volume institutions, which would reduce the average cost per patient. However, it is possible that high-volume hospitals might stimulate an unwarranted high level of healthcare utilisation, due to the easy access and the high capacity for advanced treatment and care (Fisher et al. 2000). A study from the United States found that hospital consolidation over a 17-year period was associated with increased healthcare utilisation among patients with heart disease (Hayford 2012). One interpretation of that finding might be that, when patients must travel farther for treatment, they might require more intensive treatment and it may result in higher mortality. Another interpretation could be that larger hospitals have easy access to advanced, costly equipment and procedures, and consequently, they are more likely to use these tools, though not always indicated, which would lead to inefficient care and overtreatment.

There is no conclusive evidence to demonstrate a general relationship between the patient volume and the quality of care. Furthermore, the quality of care should be balanced against other aspects of care, including patient satisfaction, the availability of healthcare and the possibility of involving caregivers.

In Nordic countries, there is an ongoing debate about the potential impacts of centralised healthcare on geographic availability and equality in healthcare. Although all the Nordic countries have brought into a common set of economic and social policies, often termed the 'Nordic model', which is characterised by a combination of free market capitalism and a comprehensive welfare state, there are also important differences between countries (Andersen et al. 2007) regarding the healthcare system. To some extent, different Nordic countries have implemented different strategies for centralising healthcare. For example, extensive centralisation has occurred at the hospital level over the last few decades in Denmark; however, to date, other countries (e.g. Norway) have insisted on a more decentralised healthcare system at both the hospital and community levels (Olejaz et al. 2012). The different strategies for centralisation that are implemented across the Nordic countries may reflect political and historical differences but also differences in geography (e.g. the length of Norway is more than fivefold larger than that of Denmark). It remains unclear what role travel distance to the nearest hospital should play in considering the most appropriate level of centralisation. Treatment delays caused by transporting patients over long distances should be weighed against the potentially more effective treatment offered at highly specialised centres. This trade-off was illustrated by the dilemma involved in selecting the most effective transportation strategy for patients with acute myocardial infarction (Andersen et al. 2003). Furthermore, although the most effective strategy can be identified for some acute medical conditions, those results are not necessarily generalisable to other disease areas. For example, in chronic conditions, where a long-term patient-centred care programme requires participation of the family, a long distance to the hospital may have potentially important adverse effects on care delivery.

FUTURE RESEARCH

Further research is clearly warranted on the differences in quality of care between high- and low-volume hospitals. Moreover, more data are required on differences in hospital organisations and the possible existence of volume thresholds. Examining the relationships between patient volume and potential mediating mechanisms will provide a better explanation of why outcome differences exist between low- and high-volume physicians and healthcare units. Advancing research in this area could also provide knowledge for improvement by identifying the factors lacking in poor-performing providers, in both low-volume and high-volume cases. As the processes of high quality are identified, it will be important to investigate how readily they can be applied to other hospital contexts. For example, some processes related to a better outcome, such as mobilising patients with hip fractures within 24 h postoperatively, can be readily shared with other hospitals. Alternatively, other processes, such as full-time geriatric service in an orthopaedic department, may be difficult for other hospitals to implement, due to an insufficient level of clinical activity to justify that level of organisation. Therefore, organisational practices should also be studied.

The volume threshold is an important area that merits further study, because hospital consolidation is likely to increase in the future. Therefore, a hospital considered large by today's standards may be considered small in the future. Continuous research will be needed to ensure that our understanding of the interplay between patient volume, quality of care and treatment outcomes remain up to date.

REFERENCES

Albornoz, C.R., Cordeiro, P.G., Mehrara, B.J., Pusic, A.L., McCarthy, C.M., Disa, J.J. and Matros, E. 2014. Economic implications of recent trends in U.S. immediate autologous breast reconstruction, *Plastic and Reconstructive Surgery*, 133(3), 463–470.

Andersen, H.R., Nielsen, T.T., Rasmussen, K., Thuesen, L., Kelbaek, H., Thayssen, P., Abildgaard, U. et al. 2003. A comparison of coronary angioplasty with fibrinolytic therapy in acute myocardial infarction, *The New England Journal of Medicine*, 349(8), 733–742.

Andersen, T.M., Holmström, B., Honkapohja, S., Korkman, S., Tson, S.H. and Vartiainen, J. 2007. *The Nordic Model: Embracing Globalization and Sharing Risks*. The Research Institute of the Finnish Economy (ETLA), Helsinki, Finland.

Auerbach, A.D., Hilton, J.F., Maselli, J., Pekow, P.S., Rothberg, M.B. and Lindenauer, P.K. 2009. Shop for quality or volume? Volume, quality, and outcomes of coronary artery bypass surgery, *Annals of Internal Medicine*, 150(10), 696–704.

Bardach, N.S., Olson, S.J., Elkins, J.S., Smith, W.S., Lawton, M.T. and Johnston, S.C. 2004. Regionalization of treatment for subarachnoid hemorrhage: A cost-utility analysis, *Circulation*, 109(18), 2207–2212.

Bates, D.W. and Gawande, A.A. 2003. Improving safety with information technology, *The New England Journal of Medicine*, 348(25), 2526–2534.

Begg, C.B., Cramer, L.D., Hoskins, W.J. and Brennan, M.F. 1998. Impact of hospital volume on operative mortality for major cancer surgery, *JAMA*, 280(20), 1747–1751.

Bennett, C.L., Garfinkle, J.B., Greenfield, S., Draper, D., Rogers, W., Mathews, C. and Kanouse, D.E. 1989. The relation between hospital experience and in-hospital mortality for patients with AIDS-related PCP, *JAMA*, 261(20), 2975–2979.

Borenstein, S.H., To, T., Wajja, A. and Langer, J.C. 2005. Effect of subspecialty training and volume on outcome after pediatric inguinal hernia repair, *Journal of Pediatric Surgery*, 40(1), 75–80.

Brook, R.H., Park, R.E., Chassin, M.R., Solomon, D.H., Keesey, J. and Kosecoff, J. 1990. Predicting the appropriate use of carotid endarterectomy, upper gastrointestinal endoscopy, and coronary angiography, *The New England Journal of Medicine*, 323(17), 1173–1177.

Creswell, J. and Abelson, R. 2013. New laws and rising costs create a surge of supersizing hospitals, *The New York Times*, p. 13.

Fisher, E.S., Wennberg, J.E., Stukel, T.A., Skinner, J.S., Sharp, S.M., Freeman, J.L. and Gittelsohn, A.M. 2000. Associations among hospital capacity, utilization, and mortality of US Medicare beneficiaries, controlling for sociodemographic factors, *Health Services Research*, 34(6), 1351–1362.

Franzo, A., Francescutti, C. and Simon, G. 2005. Risk factors correlated with post-operative mortality for hip fracture surgery in the elderly: A population-based approach, *European Journal of Epidemiology*, 20(12), 985–991.

Glassou, E.N., Pedersen, A.B. and Hansen, T.B. 2014. Risk of re-admission, reoperation, and mortality within 90 days of total hip and knee arthroplasty in fast-track departments in Denmark from 2005 to 2011, *Acta Orthopaedica*, 85(5), 493–500.

Goodney, P.P., Stukel, T.A., Lucas, F.L., Finlayson, E.V. and Birkmeyer, J.D. 2003. Hospital volume, length of stay, and readmission rates in high-risk surgery, *Annals of Surgery*, 238(2), 161–167.

Gordon, T.A., Bowman, H.M., Bass, E.B., Lillemoe, K.D., Yeo, C.J., Heitmiller, R.F., Choti, M.A., Burleyson, G.P., Hsieh, G. and Cameron, J.L. 1999. Complex gastrointestinal surgery: Impact of provider experience on clinical and economic outcomes, *Journal of the American College of Surgeons*, 189(1), 46–56.

Halm, E.A., Lee, C. and Chassin, M.R. 2002. Is volume related to outcome in health care? A systematic review and methodologic critique of the literature, *Annals of Internal Medicine*, 137(6), 511–520.

Hannan, E.L., Kilburn, H. Jr., Bernard, H., O'Donnell, J.F., Lukacik, G. and Shields, E.P. 1991. Coronary artery bypass surgery: The relationship between in hospital mortality rate and surgical volume after controlling for clinical risk factors, *Medical Care*, 29(11), 1094–1107.

Hayford, T.B. 2012. The impact of hospital mergers on treatment intensity and health outcomes, *Health Services Research*, 47(3 Pt 1), 1008–1029.

Hewitt, M.E. 2000. *Interpreting the Volume-Outcome Relationship in the Context of Health Care Quality: Workshop Summary*, Institute of Medicine, Washington, DC.

Horwitz, L.I., Lin, Z., Herrin, J., Bernheim, S., Drye, E.E., Krumholz, H.M., Hines, H.J. Jr. and Ross, J.S. 2015. Association of hospital volume with readmission rates: a retrospective cross-sectional study, *BMJ (Clinical Research ed.)*, 350, h447.

Hwang, E.W., Thomas, I.C., Cheung, R. and Backus, L.I. 2013. Assessment and utilization of rapid virologic response in US veterans with chronic hepatitis C: Evaluating provider adherence to practice guidelines, *Journal of Clinical Gastroenterology*, 47(3), 264–270.

Jha, A.K. 2015. Back to the future: Volume as a quality metric, *JAMA*, 314(3), 214–215.

Joynt, K.E., Orav, E.J. and Jha, A.K. 2011. The association between hospital volume and processes, outcomes, and costs of care for congestive heart failure, *Annals of Internal Medicine*, 154(2), 94–102.

Koelemay, M.J. and Vahl, A.C. 2007. Meta-analysis and systematic review of the relationship between volume and outcome in abdominal aortic aneurysm surgery (*Br J Surg* 94: 395–403), *The British Journal of Surgery*, 94(8), 1041; author reply 1041.

Kristensen, P.K., Thillemann, T.M. and Johnsen, S.P. 2014. Is bigger always better? A nationwide study of hip fracture unit volume, 30-day mortality, quality of in-hospital care, and length of hospital stay, *Medical Care*, 52(12), 1023–1029.

Kristensen, T., Bogetoft, P. and Pedersen, K.M. 2010, Potential gains from hospital mergers in Denmark, *Health Care Management Science*, 13(4), 334–345.

Leonard, D., Penninckx, F., Kartheuser, A., Laenen, A., Van Eycken, E. and PROCARE 2014. Effect of hospital volume on quality of care and outcome after rectal cancer surgery, *The British Journal of Surgery*, 101(11), 1475–1482.

Lindenauer, P.K., Behal, R., Murray, C.K., Nsa, W., Houck, P.M. and Bratzler, D.W. 2006. Volume, quality of care, and outcome in pneumonia, *Annals of Internal Medicine*, 144(4), 262–269.

Liu, C.J., Chou, Y.J., Teng, C.J., Lin, C.C., Lee, Y.T., Hu, Y.W., Yeh, C.M., Chen, T.J. and Huang, N. 2015. Association of surgeon volume and hospital volume with the outcome of patients receiving definitive surgery for colorectal cancer: A nationwide population-based study, *Cancer*, 121(16), 2782–2790.

Liu, J.H., Zingmond, D.S., McGory, M.L., SooHoo, N.F., Ettner, S.L., Brook, R.H. and Ko, C.Y. 2006. Disparities in the utilization of high-volume hospitals for complex surgery, *JAMA: The Journal of the American Medical Association*, 296(16), 1973–1980.

Losina, E., Walensky, R.P., Kessler, C.L., Emrani, P.S., Reichmann, W.M., Wright, E.A., Holt, H.L. et al. 2009. Cost-effectiveness of total knee arthroplasty in the United States: Patient risk and hospital volume, *Archives of Internal Medicine*, 169(12), 1113–1121; discussion 1121–1122.

Nguyen, N.T., Paya, M., Stevens, C.M., Mavandadi, S., Zainabadi, K. and Wilson, S.E. 2004. The relationship between hospital volume and outcome in bariatric surgery at academic medical centers, *Annals of Surgery*, 240(4), 586–593; discussion 593–594.

Olejaz, M., Juul Nielsen, A., Rudkjobing, A., Okkels Birk, H., Krasnik, A. and Hernandez-Quevedo, C. 2012. Denmark health system review, *Health Systems in Transition*, 14(2), i–xxii, 1–192.

Palm, H., Jacobsen, S., Krasheninnikoff, M., Foss, N.B., Kehlet, H., Gebuhr, P. and Hip Fracture Study Group. 2007. Influence of surgeon's experience and supervision on re-operation rate after hip fracture surgery, *Injury*, 38(7), 775–779.

Posnett, J. 1999. Is bigger better? Concentration in the provision of secondary care, *BMJ (Clinical Research ed.)*, 319(7216), 1063–1065.

Post, P.N., Kuijpers, M., Ebels, T. and Zijlstra, F. 2010. The relation between volume and outcome of coronary interventions: A systematic review and meta-analysis, *European Heart Journal*, 31(16), 1985–1992.

Ross, J.S., Normand, S.L., Wang, Y., Ko, D.T., Chen, J., Drye, E.E., Keenan, P.S. et al. 2010. Hospital volume and 30-day mortality for three common medical conditions, *The New England Journal of Medicine*, 362(12), 1110–1118.

Sainsbury, R., Haward, B., Rider, L., Johnston, C. and Round, C. 1995. Influence of clinician workload and patterns of treatment on survival from breast cancer, *Lancet (London, England)*, 345(8960), 1265–1270.

Slattery, W.H., Schwartz, M.S., Fisher, L.M. and Oppenheimer, M. 2004. Acoustic neuroma surgical cost and outcome by hospital volume in California, *Otolaryngology—Head and Neck Surgery: Official Journal of American Academy of Otolaryngology-Head and Neck Surgery*, 130(6), 726–735.

Svendsen, M.L., Ehlers, L.H., Ingeman, A. and Johnsen, S.P. 2012. Higher stroke unit volume associated with improved quality of early stroke care and reduced length of stay, *Stroke: A Journal of Cerebral Circulation*, 43(11), 3041–3045.

Swart, E., Vasudeva, E., Makhni, E.C., Macaulay, W. and Bozic, K.J. 2015. Dedicated perioperative hip fracture comanagement programs are cost-effective in high-volume centers: An economic analysis, *Clinical Orthopaedics and Related Research*, 474(1), 222–233.

Thiemann, D.R., Coresh, J., Oetgen, W.J. and Powe, N.R. 1999. The association between hospital volume and survival after acute myocardial infarction in elderly patients, *The New England Journal of Medicine*, 340(21), 1640–1648.

Tsai, C.L., Rowe, B.H., Cydulka, R.K. and Camargo, C.A. Jr. 2009. ED visit volume and quality of care in acute exacerbations of chronic obstructive pulmonary disease, *The American Journal of Emergency Medicine*, 27(9), 1040–1049.

Tsai, T.C., Joynt, K.E., Orav, E.J., Gawande, A.A. and Jha, A.K. 2013. Variation in surgical-readmission rates and quality of hospital care, *The New England Journal of Medicine*, 369(12), 1134–1142.

Tsugawa, Y., Kumamaru, H., Yasunaga, H., Hashimoto, H., Horiguchi, H. and Ayanian, J.Z. 2013. The association of hospital volume with mortality and costs of care for stroke in Japan, *Medical Care*, 51(9), 782–788.

Zevin, B., Aggarwal, R. and Grantcharov, T.P. 2012. Volume-outcome association in bariatric surgery: A systematic review, *Annals of Surgery*, 256(1), 60–71.

7 National Clinical Registries

Ten Years of Experience with Improving Quality of Care in Denmark

Søren Paaske Johnsen, Jan Mainz and Paul D. Bartels

CONTENTS

Information on the patterns of healthcare performance is an essential component of well-functioning and efficient healthcare systems. Healthcare data help to identify and locate

1. Problems in the provision of care
2. Regions, areas and healthcare systems that lag or lead in delivering high-quality and efficient care
3. The causes of unwarranted variation
4. The magnitude of public benefit, if problems in care were remedied

Still, most countries lack a system of healthcare surveillance that could identify what problems occur and where and that could monitor progress towards healthcare improvement (Corallo et al., 2014). Clinical registers represent a solution to these challenges as they contain systematically collected data related to clinical observations, diagnostic procedures, treatments and outcomes within the context of patient pathways of specific diseases or health-related interventions (Agency for Healthcare Research and Quality, 2014). Many countries plan to establish clinical registers and therefore look towards countries with experience within this area. The Nordic countries and, in particular, Sweden and Denmark have a long tradition for working with clinical registries and therefore have extensive experience with both the strengths and limitations of the registries (Ekman et al., 2014; Vårdanalys, 2014). In this chapter, we report on the Danish experience with using clinical registries as a central pillar in the national strategy for improving the quality of care. The authors have played an active role in establishing and running clinical registries in Denmark in the last two decades and can consequently not be considered to be neutral. However, the first-hand experience and observations made over the years may hopefully still be of value for readers with an interest in clinical registries.

SETTING AND NATIONAL PUBLIC REGISTRIES

Denmark has approximately 5.6 million inhabitants who are served by a publicly funded healthcare system granting free and equal access to visits at general practitioners, admissions and outpatient visits at hospitals for all residents (Indenrigs- og Sundhedsministeriet SFDR, 2011). The country has a long historic tradition for establishing population-based registries covering many aspects of public life.

The existence of a 10-digit personal identification number, which follows each Dane from cradle to grave and is used in all public registries, allows for unambiguous linkage between the various registries. Because of the existence of a wide range of nationwide, population-based registries, data on healthcare (e.g. contacts with general practitioners, hospital admissions and filled prescriptions at the pharmacies) can be combined with the data of a more general nature (e.g. place of living and vital status) as well as detailed socio-economic information including individual-level data on education, occupation and income.

HISTORY OF DANISH CLINICAL REGISTRIES

Clinical registries have for more than 20 years constituted a central tool for quality improvement work in the Danish healthcare system. The pioneer within this area was the Danish Breast Cancer Cooperative Group, which was established by the Danish Surgical Society in 1976 with the aim of ensuring optimal diagnostic procedures and treatment of breast cancer (Moller et al., 2008). The group has since then prepared evidence-based national guidelines regarding diagnostic procedures and treatment (surgery, radiotherapy and systemic therapy). These have continuously been developed and evaluated in randomised trials and quality control studies. Furthermore, a clinical registry was established at the same time as the guidelines. Departments of pathology, surgery and oncology have since 1976 systematically reported information about diagnostics, treatment and follow-up. Notably, these activities were initiated 10 years prior to systematic quality improvement being accepted as a necessary activity of a modern healthcare system.

The key elements in the development of the role and organisation of clinical registries in Denmark are presented in Table 7.1. The registries moved from the pioneer

TABLE 7.1
Clinical Registries in Denmark: Development 1975–2015

1975–1995	1993–1999	1999–2011	2011–2015
Bottom-up research registries	Let a thousand flowers bloom: • National clinical registers as a tool for monitoring the effect of a national quality improvement strategy + professional quality improvement/research	• National quality programme • Position paper, law and administrative order, basic requirements for clinical quality databases • National Indicator Project and three competency centres for clinical quality databases • Interdisciplinarity • Cross-sectoral registries • Involvement of organisation and management • Public disclosure of results	Universal framework and requirements for clinical registries: • One umbrella support organisation (Danish Clinical Registries) • Stable standardised reporting and public disclosure (monthly/yearly) • Uniform access for researchers • Management information • Danish Multidisciplinary Cancer Groups

phase to be a recognised element of the national quality strategy in 1993. Since then, a number of changes within and around the registries have occurred leading up to the current status. However, the fundamental principle of professional responsibility and ownership of the contents of the registries still remain as a core value in the ongoing work with the clinical registries.

CURRENT STATUS

According to Danish law, clinical registries are defined as public registries which contain individual-level information about well-defined patient groups (Sundhedsdatastyrelsen, 2016). To be approved as a clinical registry, the primary objective of the registries should be to facilitate surveillance and improvement of quality of care. However, it is also an important part of the mission of the registries to facilitate accountability and transparency in healthcare (e.g. to ensure that clinicians and administrators are accountable for the provided care and when relevant to facilitate informed patient choices regarding preferred place of care). Furthermore, there is increasing awareness of the potentially important role of the registries in the national research infrastructure (i.e. the registries represent large unselected patient cohorts with detailed clinical information and often complete follow-up, which makes them highly relevant for clinical and epidemiological research) (Table 7.2).

The registries are required to fulfil a set of national criteria regarding organisation, functionality, data safety and reporting. Once a registry is approved by the national health authority overseeing the registries (Statens Serum Insititut, 2015), reporting of all relevant patients to the database is mandatory by law for the

TABLE 7.2
Mission for Danish Clinical Registries

Improvement	Improving prevention, diagnostics, treatment and rehabilitation
Management/accountability	Documentation for clinical governance and organisational priority setting
Transparency	Information for citizens and patients
Innovation	Research infrastructure

hospitals and clinics treating the patients. The approved clinical quality databases are exempted from the requirement of obtaining the patients consent to collect relevant data.

There are currently approximately 70 approved nationwide clinical registries in Denmark, including 5 registries which are in the process of being established. The registries cover more than 80 clinical areas, with the majority covering specific disease entities. Other clinical registries cover a specific disease-related procedure (e.g. bariatric surgery or hip replacement), while a third category of databases covers a broader range of procedures such as treatment at intensive care units or anaesthetic procedures. The highest density of registries is found within cancer, major chronic diseases, orthopaedic surgery and obstetrics and gynaecology.

Most of the registries cover only secondary and tertiary care. However, collaboration with the primary healthcare sector has been established for five selected chronic diseases.

The individual registries each have a board with representation of relevant medical specialities and in many cases also other health professionals (e.g. nurses, occupational therapists, physiotherapists and dieticians). The activities of the clinical registries have since 2011 been coordinated through the umbrella organisation, the *Danish Clinical Registries* which is a national quality improvement programme responsible for the activities in the nationwide clinical registries. The registries receive support from competency centres, which provide epidemiological, IT and quality improvement support, including coordination of continuous contact between hospitals and clinics as well as the administrative and political level. The *Danish Clinical Registries* is governed by a secretariat referring to a board representing the Danish regions, health authorities, professional and patient organisations. The registries are almost exclusively publicly financed.

DATA COLLECTION AND VALIDITY

Primary collection of detailed patient-level data may require substantial time and resources, which are not always available among overloaded clinicians in busy clinical settings. Restricting the requirements for primary data collection, that is defined as data that are collected and reported specifically to the clinical registries and not otherwise used in the care of the patient, has therefore been a focus for the *Danish Clinical Registries* in recent years. The aim is to reduce the workload for the health professionals and at the same time still obtain the required data with high validity for the clinical registries. Record linkage to existing administrative and healthcare registries is therefore increasingly used as a way of re-using relevant data that are already being collected for other purposes. More than half of all variables in the clinical registries in Denmark now come from other existing public registries, and this trend is expected to continue and even accelerate in the coming years with the increasing availability of obtaining data directly from electronic medical record systems. The use of data from existing data sources has in general been successful; however, incorporating these data in the clinical registries is a demanding process, which requires substantial efforts from all stakeholders. As an example, using administrative hospital discharge data may imply challenges with a tradition of low prioritisation of correct coding and high prioritisation of diagnosis-related group coding, lack of timely reporting of data from the hospitals and time delays in processing the data in local and central administrative information systems.

A high completeness of the registration of relevant patients is a key aim in the clinical registries in order to ensure that the individual registry is representative of the entire patient population in Denmark. Consequently, a criterion for receiving legal approval and funding is that at least 90% of all relevant patients are captured by the database. This criterion is fulfilled by the large majority of the registries within 1–2 years after establishment.

In addition to the completeness of the patient registration, the completeness and quality of the recorded data for the registered patients are also essential. The clinical registries are required by the *Danish Clinical Registries* and the national health authority to have detailed data definitions and to address problems with data collection at the annual audits, including local variation in data registration practice. These approaches support a uniform and valid registration practice; however, challenges remain with the data quality in some registries and continuous attention is therefore required.

DATA REPORTING

All Danish clinical registries are required by law to produce an annual report in which the boards of the individual registries provide epidemiological and clinical interpretation of the results and make recommendations for quality improvement (Sundhedsdatastyrelsen, 2016). The reports are made publicly available at Sundhed.dk (2016). Furthermore, over the last years, a development towards more timely reporting of the data to both health professionals and administrators has accelerated following demands from the regions in Denmark, which are responsible for running the hospitals. This has been done in order to support continuous monitoring and quality of care improvement work. Results from more than 60% of the registries are now delivered monthly to the regions' management information systems, and the development has reached a point where it is the capacity of the receivers rather than the sender that restricts the access to data.

ROLE OF CLINICAL REGISTRIES IN QUALITY IMPROVEMENT

The clinical registries represent in many ways a prototype for data-driven quality improvement. Simple one-way presentations for the organisation, including management and health professionals, only rarely have effect in relation to quality improvement (Agency for Healthcare Research and Quality, 2015). If data should have an effect, they must be selected in a way that they are both relevant and valid. This is done by the boards of the individual clinical registries, which represent the practical experience in the clinical field supplemented by high-level expertise in clinical epidemiology, biostatistics and data management ensuring credible methodology.

In relation to the clinical registries in Denmark, it has been our experience that it is an important strength that members of the boards of the clinical registries are strongly involved in the writing of clinical guidelines and the construction of clinical pathways. This ensures that the clinical registries and the clinical guidelines mutually support each other and that the registries can be used as a tool for supporting the implementation of the recommendations in the clinical guidelines.

There is currently an increasing recognition of the importance of the context for quality improvement based on quality indicator data. The detailed data of the clinical registries in relation to both the patients (e.g. the severity of the disease of the individual patient) and the provided care (e.g. details about the procedures and treatments) make it possible to ensure that the data are both meaningful and credible in different contexts. This has ensured clinical ownership and facilitated a reflective use of the data.

Substantial improvements in the quality of care have been observed, in particular within the last decade, for the majority of the medical areas in Denmark covered by clinical registries. The improvements include both processes of care and clinical outcomes (Møller et al., 2013; Nakano et al., 2013; Rosenstock et al., 2013; Tøttenborg et al., 2013; Jørgensen et al., 2015). A causal association between the quality improvement work driven by the data from the clinical registries and the observed improvements in care is difficult to document; however, given that the indicator monitoring and systematic auditing based on the clinical registries has been the most consistent quality improvement activity in Denmark during the period, it is probably reasonable to assume that the clinical registries have played an important role for the changes in care that have occurred over the years.

TARGETS FOR QUALITY OF CARE

A particular important factor making the observed improvements in care possibly appear to have been the increasing involvement of the entire organisation in using the data from the clinical registries for managerial support for professional-led quality improvement. This involvement has to a large extent been based on targets or standards for the quality of care, which has been defined by the health professionals. An example could be that at least 90% of the patients with acute stroke should be admitted or transferred to a specialised stroke unit or that at least 90% of the patients with systolic heart failure should receive an angiotensin-converting enzyme inhibitor or angiotensin II blocker. This is from an international perspective a unique feature of the Danish clinical registries (Agency for Healthcare Reaserch and Quality, 2014; Ekman, 2014). Target values are defined via available evidence or in the absence of this via a formalised process of expert consensus. Initially, the target construction was described in 1999 by an expert group commissioned by the Danish medical societies and hospital owners and implemented in the Danish Indicator Project. In 2005, the concept of indicator plus standard was codified in the approval form issued by the National Board of Health for mandating Danish clinical registries.

The targets have in many ways been successful as aids for improvement, governance and accountability by national clinical registries in Denmark. Since they are, at least in principle, legitimated by science and professionals, they have been accepted throughout the healthcare system making it easier for the organisations and the surrounding society to identify areas where improvements are warranted. Although the approach has had a positive impact, in particular on structural changes in the healthcare system (e.g. establishment of specialised stroke units and heart failure clinics), it also appears to be areas where the original intention of the targets has not been followed. This development reflects that the demonstration of actual evidence for best practice coupled with the standards is quite weak for many medical conditions. Furthermore, the professional definition of standards for best practice has been highly variable ranging from ideal goals to thresholds ranging from acceptable to unacceptable. This variation obviously has major implications for the interpretation of the targets. However, it is noteworthy that perception of the standards at the management level appears to have been consistent across the clinical registries (i.e. the targets have all been interpreted as evidence based and reflecting ideal goals). These discrepancies have to be resolved if the targets shall continue to be a tool for quality improvement and governance. A simple solution would be to ensure that the targets are classified according to their context. Thus, we need to clearly distinguish between best practice internationally, nationally derived targets and targets reflecting acceptability.

CLINICAL REGISTRIES AND RESEARCH

The first generation of clinical registries in Denmark was primarily established for research purposes, and clinical motivation and acceptance for participation in the work with the clinical registries is still to some extent conditioned by the possibility of using the collected data for research purposes. Although the involvement of other stakeholders in the clinical registries has increased dramatically over the years and has had major impact on the mission of the registries, the academic activities remain essential for the development of the registries.

The clinical registries come with a number of methodological strengths that make them viable from a research perspective. First, they make it possible to conduct population-based studies on large cohorts of patients with usually close to complete data of the study population including follow-up. Additional advantages include the fact that data are collected independently of any research question, which reduces the risk of information bias, for example recall bias and impact of decision on diagnosis and therapy because of the awareness of an ongoing study. The fact that data are already collected also means that the cost of research in clinical quality databases is usually low compared with clinical studies based on primary data collection. However, limitations still exist and should be acknowledged. The data collection is not controlled by the researcher, and patient characteristics may be poorly measured or even lacking. Still, most clinical registries contain more detailed clinical data than alternative data sources, in particular administrative health registries. In addition, data from the clinical registries can be complemented through linkage with other registries and data from medical records.

During the last decade, data from the Danish clinical registries have been used in a wide range of research. This includes traditional clinical epidemiological studies on patient prognosis focusing on determining the outcome of specific groups of patients, identification of prognostic factors or prognostic models. Another group of studies have focused on medical care variation and have tried to identify and explore unwarranted variation in medical care defined as differences in the performance of healthcare systems, rather than in population needs or preferences (Corallo, 2014). Examples have been studies on disparities in quality of care according to age, sex, socio-economic status or time of admission. Another important research area has been comparative effectiveness research, which can be defined as direct comparison of existing healthcare interventions to determine which works best for which patients and which pose the greatest benefits and harms in routine clinical settings (Sullivan and Goldmann, 2011). As an example, the Western Denmark Heart Registry has been used as platform for a series of pragmatic clinical trials comparing different types of coronary stents in patients with coronary heart disease (Christiansen et al., 2013; Maeng et al., 2014). In addition, the registries have also provided data for health delivery research, which is also known as translation research or implementation science (Pronovost and Goeschel, 2011).

Finally, the clinical registries have participated in international studies, in particular collaboration studies between the Nordic countries. This type of collaboration is facilitated by the close similarities of the political systems in the Nordic countries and the comprehensive tax-financed healthcare systems covering entire populations. As an example, Mäkelä et al. (2014) used data from the Nordic Arthroplasty Register Association which combine data from Sweden, Norway, Finland and Denmark to demonstrate that the survival of cemented implants for total hip replacement was higher than that of uncemented implants in patients aged 65 years or older.

FUTURE PERSPECTIVES

When the clinical registries have been able to continuously contribute to the development of patient care in the Danish healthcare system over a long time period, it is to a large extent due to an innovative clinical practice that has taken advantage of existing trends (technological, organisational and attitudinal) in the healthcare system. Throughout the last decades, there have been several major changes in the strategy for the clinical registries. These changes have been driven both by clinicians and by other stakeholders. The last major change occurred in the period between 2009 and 2011, but new changes are already underway. The most important element is an increasing recognition and acceptance of the need to give more priority to direct patient involvement in the management of the healthcare system. This is particularly important within the area of quality of care, which in Denmark so far has been very focused on the formal role of the involved stakeholders. However, there are already experiences with clinical registries that have been proactive in taking patient preferences and needs into account in relation to data-driven quality improvement (e.g. the Danish Biologics database (DANBIO) – a national registry for rheumatologic patients treated with biological drugs).

It has already been demonstrated that it is possible to integrate patient-reported data on outcomes and experiences into quality of care work based on clinical registries (Baker et al., 2012). In addition, there are positive experiences from Sweden with involvement of patients in the selection of relevant areas for quality improvement, interpretation and communication of data (Ekman, 2014). This means that one of the key focus areas for the Danish clinical registries will be increased involvement of patients in prioritisation and interpretation of the quality of care data as well as large-scale use of patient-reported data within relevant areas.

Another focus area will be timely reporting of data to ensure that clinical registries can facilitate a dynamic and ongoing clinical practice in which data-driven quality improvement work is an integrated part of everyday practice and not a once-a-year procedure aimed at retrospectively explaining data. Targeted communication aimed at the individual stakeholders (general population, patients, decision-makers, management, health professionals) will be important.

The current clinical registries cover very different areas (e.g. from the quality of care of a specific procedure to lifelong courses of chronic diseases, which covers numerous hospital departments and providers in the healthcare system). The current technological development in relation to the electronic medical records will imply that 'simple' surveillance of specific procedures can be monitored directly using the electronic medical record. This will require that the electronic medical record can include all data elements that form the basis for the registry and hereafter report directly to the locally involved clinicians. It is therefore expected that it will be possible to monitor these areas with less resources in the future. In contrast, the complex courses of diseases experienced by patients with cancer and chronic diseases will continue to require the full set-up of a clinical registry, and these areas will consequently be given high priority.

Finally, there is a need for the clinical registries to play an even more active role in the generation of new knowledge about the performance of healthcare system. Although there have been examples of health services research studies based on clinical registry data, much more is needed in order to take advantage of the information potential, which is currently not exploited given the substantial human and financial resources invested in the clinical registries. A more active and systematic approach is therefore needed in order to ensure that important differences in the quality of care are revealed, the causes of unwarranted variation are identified and programmes intended to improve healthcare are evaluated. To obtain a broader picture, analyses across clinical registries will be needed.

It is expected that the clinical registries by measuring avoidable variation may help identifying misused resources that can be reallocated to achieve better value without sacrificing equity. Better public transparency and research on variation in healthcare performance may potentially accelerate improvements in care with benefits to patients while reducing costs. In contrast, the absence of data will blind patients, physicians, healthcare managers and policymakers from the information needed to understand and improve care. Data gathering, data access and independent data analysis are essential to the national provision of high-value healthcare.

Initiatives are currently underway to further strengthen the role of the clinical registries in the national research infrastructure in Denmark. The Program for Clinical Research Infrastructure is a national initiative aiming at improving integration across clinical registries, other public administrative and healthcare registries and biobanks in the healthcare sector to facilitate data analysis and to integrate research findings into daily clinical work by building bridges between research and clinical practice. Focus will also be on the development of novel methods for analysing large data sets due to the growing amount and complexity of the information available within healthcare (e.g. medical and administrative databases, electronic medical records and biobanks).

CONCLUSION

In conclusion, the Danish clinical registries have played a central role in the data-driven quality improvement work in Denmark over the last decades and in particular within the last decade. The key element in this development has been an early decision to make clinical registries the backbone of the national quality improvement strategy and subsequently to involve all stakeholders in the work. The efforts have been strongly facilitated by legislation, which regulates the registries and ensures that requirements for institutions and hospital owners, in particular mandatory reporting of all relevant patients to the registries, are given high priority. Furthermore, transparency and accountability through target setting and public disclosure of all performance data at the unit level have stimulated professional engagement and interest. Finally, technical aspects including a strategy to ensure uniform software solutions and standardisation of the formats for output/reporting have been important. The Danish clinical registries face a number of still unsolved challenges; however, the elements mentioned earlier will most likely remain central also in the years ahead.

REFERENCES

Baker, P.N., Deehan, D.J., Lees, D., Jameson, S., Avery, P.J., Gregg, P.J. and Reed, M.R. 2012, The effect of surgical factors on early patient-reported outcome measures (PROMS) following total knee replacement. *J Bone Joint Surg Br*, 94(8): 1058–1066.

Christiansen, E.H., Jensen, L.O., Thayssen, P., Tilsted, H.H., Krusell, L.R., Hansen, K.N. et al. 2013, Biolimus-eluting biodegradable polymer-coated stent versus durable polymer-coated sirolimus-eluting stent in unselected patients receiving percutaneous coronary intervention (SORT OUT V): A randomised non-inferiority trial. *Lancet*, 381(9867): 661–669.

Corallo, A.N., Croxford, R., Goodman, D.C., Bryan, E.L., Srivastava, D. and Stukel, T.A. 2014, A systematic review of medical practice variation in OECD countries, *Health Policy*, 114: 5–14.

Ekman, G.J., Lindahl B. and Nordin, A. (eds). 2014, *Nationella kvalitetsregister i hälso- och sjukvården*. Stockholm, Sweden: Karolinska Institutet University Press.

Gliklich, R., Dreyer, N., Leavy, M. eds. 2014. *Registries for Evaluating Patient Outcomes: A User's Guide*, 3rd edn. (Prepared by the Outcome DEcIDE Center [Outcome Sciences, Inc., a Quintiles company] under Contract No. 290 2005 00351 TO7.) AHRQ Publication No. 13(14)-EHC111. Rockville, MD: Agency for Healthcare Research and Quality. http://www.effectivehealthcare.ahrq.gov/registries-guide-3.cfm.

Indenrigs-og Sundhedsministeriet SFDR. 2011, *Kvalitetsoplysninger på sundhedsområdet*. Copenhagen, Denmark: Indenrigs- og Sundhedsministeriet.

Jørgensen, M., Mainz, J., Svendsen, M.L., Nordentoft, M., Voldsgaard, I., Baandrup, L., Bartels, P. and Johnsen, S.P. 2015, Improving quality of care among patients hospitalised with schizophrenia: A nationwide initiative. *Br J Psychiatry Open*, 1(1): 48–53.

Maeng, M., Tilsted, H.H., Jensen, L.O., Krusell, L.R., Kaltoft, A., Kelbaek, H. et al. 2014, Differential clinical outcomes after 1 year versus 5 years in a randomised comparison of zotarolimus-eluting and sirolimus-eluting coronary stents (the SORT OUT III study): A multicentre, open-label, randomised superiority trial. *Lancet*, 383(9934): 2047–2056.

Makela, K.T., Matilainen, M., Pulkkinen, P., Fenstad, A.M., Havelin, L., Engesaeter, L. et al. 2014, Failure rate of cemented and uncemented total hip replacements: Register study of combined Nordic database of four nations. *BMJ*, 348: f7592.

Møller, M.H., Larsson, H.J., Rosenstock, S., Jørgensen, H., Johnsen, S.P., Madsen, A.H. et al. 2013, Quality-of-care initiative in patients treated surgically for perforated peptic ulcer. *Br J Surg*, 100(4): 543–552.

Moller, S., Jensen, M.B., Ejlertsen, B., Bjerre, K.D., Larsen, M., Hansen, H.B. et al. 2008, The clinical database and the treatment guidelines of the Danish Breast Cancer Cooperative Group (DBCG); its 30-years experience and future promise. *Acta Oncol*, 47(4): 506–524.

Nakano, A., Johnsen, S.P., Frederiksen, B.L., Svendsen, M.L., Agger, C., Schjødt, I. et al. 2013, Trends in quality of care among patients with incident heart failure in Denmark 2003–2010: A nationwide cohort study. *BMC Health Serv Res*, 5(13): 391.

Pronovost, P.J. and Goeschel, C.A. 2011, Time to take health delivery research seriously. *JAMA*, 306(3): 310–311.

Rosenstock, S.J., Møller, M.H., Larsson, H., Johnsen, S.P., Madsen, A.H., Bendix, J. et al. 2013, Improving quality of care in peptic ulcer bleeding: Nationwide cohort study of 13,498 consecutive patients in the Danish Clinical Register of Emergency Surgery. *Am J Gastroenterol*, 108(9): 1449–1457.

Sundhed.dk, 2016. Yearly reports from clinical registries. https://www.sundhed.dk/sundhedsfaglig/kvalitet/. Accessed 17 April, 2016.

Sundhedsdatastyrelsen. Regionernes kliniske kvalitetsdatabaser. http://sundhedsdatastyrelsen.dk/da/registre-og-services/om-de-kliniske-kvalitetsdatabaser. Accessed 17 April, 2016.

Sullivan, P. and Goldmann, D. 2011, The promise of comparative effectiveness research. *JAMA*, 305(4): 400–401.

Tøttenborg, S.S., Thomsen, R.W., Nielsen, H., Johnsen, S.P., Frausing Hansen, E. and Lange, P. 2013, Improving quality of care among COPD outpatients in Denmark 2008–2011. *Clin Respir J*, 7(4): 319–327.

Vårdanalys, 2014. Registrera flera eller analysera mera? Delutvärdering av satsningen på nationella kvalitetsregister. Rapport 9. Stockholm, Myndigheten för vårdanalys.

8 Side Effects of Overdoing It

Lessons from a Comprehensive Hospital Accreditation Programme

Carsten Engel and Henning Boje Andersen

CONTENTS

INTRODUCTION

Healthcare accreditation programmes have been developed and implemented in many countries based on the expectation that they will improve patient safety and quality of healthcare. Accreditation is not a new phenomenon, having a history of almost 100 years in the United States and Canada and was launched in Australia more than 40 years ago (Scrivens, 1995). While it maintains its importance in these countries, a steep increase in the number of accreditation programmes in operation began in the 1990s and amounted to 44 in 2009, some of which were operating in several countries (Shaw et al., 2013). In Denmark, accreditation was adopted as a national strategy in 2005 but has been partially discontinued in 2015 (see Box 8.1). In the other Nordic countries, it has never gained a foothold, although there are isolated examples of hospitals choosing to go for accreditation or ISO certification (e.g., Hasman, 2012; Lie and Bjørnstad, 2015; Norén and Ranerup, 2015).

BOX 8.1 THE DANISH HEALTHCARE QUALITY PROGRAMME

The Danish Healthcare Quality Programme (DDKM – Den Danske Kvalitetsmodel) was established in 2005 by the Danish Government and the five Danish regions (the regions run about 98% of all hospital services, financed over state taxes). The intention was to create a common national system for the assessment of quality in healthcare to promote better patient pathways, transparency of quality and continuous quality improvement (QI). At the outset, the programme covered only public hospitals, but step by step all publically financed healthcare has been included and now covers private hospitals, pharmacies, prehospital care, some aspects of municipality-based primary care as well as, from 2015 and onwards, programmes for all practice-based healthcare (GPs, physio-/ergotherapists, chiropractors) are being launched.

The two main components of the programme are the accreditation standards, setting the criteria against which the performance of the hospitals is assessed and the external survey conducted once every 3 years, where a team of surveyors assess the hospitals against the standards. The surveyors are hospital employees (senior clinicians or managers), trained for the purpose, and working as surveyors for 2–3 weeks per year. A large hospital will be assessed by a team of 6–8 surveyors for 5 days. The focus, in particular in the second cycle, has been on implementation of good working practices, judged by interviewing front-line staff and cross checking with documentation, such as patient records, and on a systematic approach to QI. Processes are assessed for their ability to deliver the expected outcome, rather than merely on compliance with formal specifications in written policies. The findings during the survey are presented in a published report and are the basis for the decision of which level of accreditation to award.

By the end of 2015, two cycles of accreditation had been completed for all public hospitals in Denmark and the programme for public hospitals was terminated. The Ministry of Health has indicated that the two cycles of accreditation have improved quality, especially organisational quality, and patient safety, but that the time has come to replace accreditation with other means of QI. Hospital accreditation is now perceived as entailing a burden of bureaucracy, excessive registration and documentation and a focus on detailed specification of processes, all of which are not meaningful to front-line staff (Danish Ministry of Health, 2015). Accreditation programmes will continue in private hospitals and pharmacies, GP practices and specialist physician practices.

The International Society for Quality in Health Care proposed that accreditation may be defined as follows:

> An act of granting credit or recognition by an external evaluation organisation of the achievement of accreditation standards, demonstrated through an independent external peer assessment of that organisation's level of performance in relation to the standards.
>
> **ISQua (2013a)**

While the intentions behind the considerable efforts involved in implementing and running accreditation programmes certainly are based on a desire to improve safety and quality, the evidence for actual improvement is scant or non-existent (Hinchcliffe et al., 2012; Brubakk et al., 2015). Still, several studies (Braithwaite et al., 2010; Schmaltz et al., 2011; Falstie-Jensen et al., 2015a,b; Bogh et al., 2015) have addressed the link between accreditation and safety and quality, but we wish to focus on another issue, namely the side effects of accreditation in terms of overimplementation of standards and the drift towards bureaucratisation. This has been elucidated in investigations of professionals' attitudes (Verstraete et al., 1998; Paccioni et al., 2008; Hinchcliffe et al., 2012), and we wish to dig deeper into this problem in this chapter.

An important background for our discussion is the well-known difficulties in designing studies to assess the effectiveness of complex interventions (Craig et al., 2008). Accreditation is clearly not a standardised intervention; there are important differences between accreditation programmes, and not least between the contexts in which they operate and in the ways in which they are understood and implemented by the organisations undergoing accreditation. Thus, it is doubtful if the findings of even a well-designed randomised controlled study could be translated to other accreditation programmes in other settings. To understand the usefulness and challenges in accreditation we must explore the mechanisms by which it is supposed to exert a beneficial effect; this might be formulated as a programme theory (Davidoff et al., 2015) – see also Chapter 4. The value of a programme theory is not only in evaluation but also in implementation. Improving the function of complex systems is a complex venture. If based on a too simple and inexplicit theory, side effects may occur that might have been avoided, had the expected mechanisms and causal relations been spelled out in advance.

One key assumption behind accreditation programmes is that standardisation of procedures will improve quality and safety (https://www.accreditation.ca/why-accreditation-matters; Roberts et al., 1987; Shaw, 2000; Rozich et al., 2004). This assumption is plausible, but we argue that this hypothesis can be interpreted too simplistically and that standardisation by way of accreditation is liable to bureaucratisation and staff resistance. We will explore our claim by presenting a brief outline of a more refined programme theory, based on the science of complex systems, and then compare this to the assumptions that have determined the actual implementation of an accreditation programme. We will do so by reviewing the comprehensive, nation-wide accreditation programme for public hospitals that ran in Denmark for 10 years until 2015. In our review, we will mainly focus on the perceptions and experience by hospital staff and healthcare decision and therefore also on the reasons for the eventual dismantlement of the hospital accreditation programme (see Box 8.1).

When the Danish accreditation initiative was launched for the Danish hospitals in 2008 (the Danish Healthcare Quality Programme, see Box 8.1), the then Managing Director Europe for Joint Commission International, Carlo Ramponi, offered this warning:

> If the bureaucratic approach is prevailing, much energy will be used to demonstrate formal compliance with standards, but there will be no real willingness to change and improve patient safety.
>
> **Mandag Morgen (2008)**

It is this drift towards bureaucratisation and excessive weight on documenting and demonstrating compliance – which can be summarised as 'overimplementation' – that we will review and discuss in this chapter. Our aim is, first, to document and describe how the original aims of the accreditation programme became overshadowed by bureaucratisation and compliance seeking – as perceived by key stakeholders and in particular clinical staff – and, second, to discuss how the drift towards bureaucratisation of QI programmes may be curtailed. In short, we seek to draw out lessons learned from the 10-year old and now terminated history of the Danish Healthcare Quality Programme for public hospitals.

INTERVIEW SURVEY OF PRACTITIONERS' EXPERIENCE AND VIEWS OF ACCREDITATION

The IKAS institute (Institut for Kvalitet og Akkreditering i Sundhedsvæsenet) is responsible for developing and managing the Danish Healthcare Quality Programme (see Box 8.1) and carried out a series of 17 focus group interviews, May to September 2014, in order to gather users' and other key stakeholders' requirements and wishes to prepare version 3 of the healthcare quality programme for hospitals. Thus, the interviews were an opportunity for eliciting experiences (positive and negative) of clinical staff with the programme including the way in which it had been implemented.

The IKAS interviews covered 12 focus group interviews with hospital and hospital quality managers on several levels (including clinical leaders) across the country and 5 with healthcare safety and quality societies and public agencies. The total number of informants amounted to 140. Results of the interviews are described in a short report published January 2015 (IKAS, 2015b). The study shows that participants and especially clinical staff have a number of critical views of the programme, but the summaries of interviews do not show a uniformly negative picture, and it should be kept in mind that the emphasis was on features that participants would like to see changed in the planned subsequent version of the Programme.

The standard set is divided into 82 *standards*, each of which describes requirement for a particular activity, such as quality surveillance or safe surgery. Compliance is assessed by 471 measurable elements, in DDKM called indicators. Other accreditation programmes use a number of different terms for the DDKM indicators, such as measurable elements or criteria.* Indicators are also used by hospitals implementing DDKM to operationalise the content of the standards, although a self-assessment is expected to reveal that the hospital is in compliance with most indicators so that action is only required for a subset. Two areas of concern that were addressed in interviews are of particular interest for the themes of this chapter, namely the nature of *standards* contained in the programme and the *implementation* of the standards, that is the selection of indicators and derived requirements to, for instance, documentation.

* http://www.achs.org.au/field-review-of-the-draft-equip6-standards,-criteria-and-elements/; http://www.jointcommissioninternational.org/jci-accreditation-standards-for-primary-care-centers/.

While hospital managers stated that in their view the programme has had a very beneficial influence on quality and safety, most other interviewees including, in particular, clinical staff shared a critical view on several aspects of the programme *standards*. In summary, in their view, the standards:

- Are in a number of instances too far removed from daily clinical work
- Have too much focus on the organisation, infrastructure and the hospital operation
- Do not clearly distinguish between requirements that are dictated by legislation and those that are imposed on the basis of quality concerns
- Are too static and do not seem to allow that standards, once their fulfilment has been demonstrated, may no longer require a continuous level of organisational attention
- Fail to include the quality assurance departments themselves, whose activities have a significant influence on the conditions under which the clinical departments can work

Similarly, participants had critical views on the *implementation* of standards – in contrast to the *contents* of the standards as earlier mentioned, and they are particularly concerned with the tendency to go beyond the requirement of the standards and engage in what we characterise as 'overimplementation'. A large proportion of participants hold that implementation:

- Has been appropriated and more or less taken over by the central regional and the hospitals quality departments with little involvement of clinical departments
- Sets the bar much higher in terms of the level, detail and scope of indicators than is justified by the contents of the standard
- Led to a proliferation of guidelines that inflate the intentions behind the quality programme
- Has led to an emphasis on passing the accreditation 'exam' with no critical comments at all, that is with no recommendations for improvement
- Has led to the use of performance measures that some participants refer to as a 'control culture' and a 'checklist culture'
- Has failed to involve sufficiently clinical department in defining indicators and their level and scope and hence their involvement in interpreting the standards in a clinically meaningful way

In summary, key stakeholders find that standards do not sufficiently reflect the clinical work context and that current practices of demonstrating and verifying compliance have been captured by an over-reliance on documentation and a search for compliance errors.

Results similar to those of the IKAS survey were also uncovered in another and independent survey carried out in 2014–2015 by the Danish Institute for Local and Regional Government Research and published in a comprehensive report in May 2015 (Holm-Petersen et al., 2015).

The results of the surveys did not come as a surprise. In recent years, a growing and often very explicit disenchantment with 'documentation load' has been expressed and described in the medical weeklies and in the general media and has been a theme for heated debate at meetings in healthcare quality and safety societies. Thus, a campaign to recruit support against growing demands for documentation in 2011 managed to collect signatures among a fifth of all hospital physicians in the capital and received extensive press coverage (Rasmussen, 2011).

OVERIMPLEMENTATION OF STANDARDS

As we have seen, hospital staff have experienced an exaggerated emphasis on documentation and in general overimplementation of standards. An important source of overimplementation, as suggested in the quote in the first section by Carlo Ramponi, is a shift in the mind of the management of the organisation from identification of opportunities for improvement to demonstration of compliance with standards ('obtaining a clean certificate'). Instead of seeing the accreditation survey as an opportunity to gain new insight into the modus operandi of the organisation, it becomes important to managers to guard themselves against surprises and to assure a satisfactory outcome of the award of status. To achieve this, a quality bureaucracy is set up and made accountable for delivering unblemished accreditation. To match expectations, the 'qualitocrats' need to be assured of process compliance. Thus, while satisfactory result of the assessment a survey should obviate the need for continuous supervision by creating confidence in the ability of the processes to perform as desired, instead, paradoxically, a tight system of internal control is established, associated with extensive demands for documentation, including documentation of non-actions.

In addition to the misunderstanding of the intention to standardise, assuming that the aim is to 'standardise everything', you arrive at what might be called the perverted quality cycle:

- Describe in details how everything should be done.
- Require everyone to document every action in details.
- Check if documented actions correspond to prescribed actions.
- Impose sanction if this is not the case.

We will further develop this proposed explanation (ex post facto) of the bureaucratisation of the use of standards by formulating a sketch of a programme theory of accreditation and comparing with examples of accreditation gone awry, taken from experiences with DDKM.

Before we describe the sources behind the possible defences against overimplementation, it will be helpful to consider the role of accreditation in a strategy for QI.

OUTLINE OF A PROGRAMME THEORY FOR ACCREDITATION

While the promotion of QI is an explicitly stated purpose of accreditation (ISQua, 2015), accreditation should not be viewed as a complete QI intervention but as an evaluation tool, used by the organisation itself and by external evaluators. Through evaluation, the organisation identifies opportunities for improvement and is thereby provided with a direction for the QI efforts of the organisation. Thus, accreditation is a component of a QI strategy, but the total strategy is more than a set of accreditation standards and an assessment methodology. And very importantly, the strategy can usually not be determined and controlled entirely by the accreditor. As for any approach to QI, the way in which the organisation imbeds accreditation in its total management strategy is of paramount importance (Dixon-Woods et al., 2014).

The assessment inherent in accreditation results not only in a set of suggested improvements but it also entails that when an accreditation is awarded it must be reviewed at regular intervals (ISQua, 2013a,b). Failure to achieve the desired status can have serious consequences for an organisation. Accreditation and other forms of external assessment can be seen as an external pressure towards QI. While positive associations between external assessment and hospital outputs can be demonstrated (Sunol et al., 2009), external pressure to obtain accreditation can also have a negative impact. This is not a new insight (McMillen et al., 2008).

The key element of a programme theory for accreditation is in our context that healthcare organisations and systems should be considered as complex systems (Dekker, 2011). This is a short statement with profound implications. A complex system cannot be totally predetermined by procedures and to accommodate unforeseen variations in context and individual differences judgement must be exercised (see also Wears, 2015a,b).

A programme theory should address in more details the way in which the organisation manages quality and risk. Suffice to say that we do not think that quality management will include detailed and continuous documentation and monitoring of compliance with process specifications. Also, the role of the external evaluation component should be clarified: What is evaluated? By what means? What kind of credit will it provide? How will it compare to other ways of identifying opportunities for improvement? (Cf. the definition of accreditation in ISQua, 2013a).

Standardisation is thus one of the several mechanisms used in accreditation. However, while accreditation standards when used as evaluation tools may reveal a need for standardisation, they should not be translated directly into written operational procedures without due consideration of what should be standardised.

How this works in practice can be illustrated with some examples, taken from our experience with the Danish Healthcare Quality Programme.

Example 8.1: Mistaking an Artefact for a Goal: Focus Shift from Outcome to Poorly Chosen Performance Indicator

Accreditation standards may include a criterion such as 'Results of diagnostic test (lab, x-ray, etc.) should be signed off'. The meaning of this is not that a signature on a piece of paper or in an electronic patient record makes any real difference. The criterion is shorthand for 'There should be a process in place to ensure that test results are seen by a competent person in a timely manner, are acted upon appropriately, and are then signed off. When a signature is missing this is then liable to alert another person to the fact that the test result has not been seen'. This is a process well suited for standardisation, although even with a process as generic as this, one should not assume that one process description can fit everywhere in a complex healthcare provider organisation. But the real caveat is that you should not focus too much on the readily observable artefact, the signature. Accreditors should evaluate compliance by checking by interviews if there is a common understanding of the work process among involved workers. Some patient records should be checked as verification, but the assessors should not hunt for an example of a missed signature. Even more importantly, the organisation should be very careful to avoid collecting data on the presence of signatures. This may shift focus from acting timely to getting the signature right, leading to absurd undertakings such as having junior doctors signing off old and now obsolete test results in assembly line fashion. The expected outcome is the absence of events due to missed test results, but measuring outcome by counting non-occurrences of rare events is problematic. If there is a perceived problem, time might be better spent doing an in-depth analysis of the workflow, to understand why it works most of the time, but still fails too often (Hollnagel, 2014). As part of an improvement project it may be warranted to track the frequency of signing for a limited period.

The example can be generalised: use of checklists promotes safety in certain circumstances and as part of a more comprehensive intervention (Dixon-Woods et al., 2011), but monitoring quality by monitoring documentation for completed checks will in most cases not be useful and will alienate professionals.

Example 8.2: Standardisation on the Wrong Level

Our second example illustrates how standardisation becomes overimplemented by being taken it too far. It is well documented that malnutrition is not rare among certain patient groups (Kondrup et al., 2002). It therefore makes good sense to screen patients at risk for malnutrition, and simple tools for this are available (Danish Health and Medicines Authority, 2008). One such tool entails asking four simple questions; a scoring matrix then enables one to decide whether there is reason to initiate further investigations, or if it can safely be assumed that the risk is minimal and no further action required.

The apparent simplicity of the screening procedure has led some quality managers and some healthcare professionals with a special interest in nutrition to suggest that every patient should be screened. We note first that this is in contrast to the logic in other types of screening; for example, while it is clearly possible that a woman aged 25 years has a yet undetected mammary cancer, screening is reserved for much older women, where the risk is considerably higher. Second, we observe that is not that easy to determine what we mean by 'every patient'. Should we screen at every patient admission? Should we also screen patients visiting outpatient clinics? Some of them might certainly be at risk. Should we screen everyone once or every week or month? If so, how do we keep track of who needs a screening? Maybe we need not screen everyone. It is stated in the Danish accreditation standards for hospitals that nutritional screening is for patient groups with a potential nutritional risk (IKAS, 2012). But if we try to make an exhaustive definition of who these patients might be, that is standardising on total patient population level, who should be screened, then our best guess is that this can never be defined satisfactorily, and it might become more cumbersome to determine whether a screening is needed than performing the actual screening. If on top of this we add mistaking means for the end and want to monitor whether we have actually screened as prescribed, then the staff may face a requirement to document not only screenings that have been performed but also whenever it has been assessed that a nutritional screening is *not* needed.

If instead we standardise at the patient pathway level, things look very different. Now, we see nutritional screening as a standardised building block: Whenever we see a need to do a nutritional screening, we perform and document it in a standardised way. And whenever we design a patient pathway, we consider whether nutritional screening should be part of it. This approach emphasises that nutritional care is not an isolated fragment but an integral part of total care.

Incidentally, patient pathways illustrate the principle that in some cases standardisation does not provide us with a final answer, but a common point from which to set off. Not all patient journeys will fit into standardised patient pathways. But whenever nutrition is deemed to be an issue by the relevant clinician, a standardised screening tool is available for use.

We should emphasise that the intention is not that screening again should be left to the discretion or the memory of the individual clinician. But standardisation should be at the level of specific patient pathways. It may require an effort to make healthcare professionals aware that the risk for malnutrition is one of the many issues they should consider, when designing a pathway or facing a patient not captured by a specific pathway, and it may be that a clever design of an electronic patient record, making a screening tool readily available, would promote this awareness.

DISCUSSION

One might wonder why overimplementation occurs. In his analysis of how evaluation is used in the modern society, Dahler-Larsen (2012) demonstrates a series of three socio-historic epochs: the modern society, the reflexively modern society and the audit society. The latter is characterised by being obsessed with a desire to verify and check in order to avoid risk and anything falling below standards. This leads to the rise of evaluation machines: 'Mandatory procedures for automated and detailed surveillance that give an overview of organisational activities by means of documentation and intense data concentration' (op. cit., p. 176). Accreditation can easily be made to fit into this image, and this will then generate the very examples of excessive procedures illustrated earlier. The general trend towards the audit society, which manifests itself in a multitude of other evaluation machines, will ensure that accreditation drifts in this direction, unless actively countered. Similarly, in his comprehensive review of the bureaucratisation of safety, Dekker (2014) describes and documents from a large and varied set of studies how the formalisation of safety standards has led to serious negative side effects, in particular a one-sided focus on compliance and a control culture that hampers innovation in QI efforts.

What could accreditors do to counteract the tendency to overimplementation?
There is an inherent tension in a system aimed both at demonstrating opportunities for improvement to an organisation and at issuing a certificate to provide

assurance to commissioners and service users. This is reinforced by the fact that the accreditation process in many jurisdictions, including Denmark, is completely transparent, whereas traditionally the detailed report and a failure to obtain accreditation have been confidential information. Transparency is a condition in a publically financed healthcare system. There is another handle, though. Nowadays, the external assessment is the final step after a prolonged phase of preparations. This seems to be contradictory to viewing the assessment as a starting point for improvement activities.

There are a number of misconceptions that the accreditors could and should address clearly.

First, standards should not be considered as rules such as, for instance, building regulations. It is not that every standard or – every measurable element – requires a specific action. The frequent use of the phrase 'implementing standards' may actually inadvertently induce this misperception and should possibly be avoided. The standards should be seen as a guidance, that is an opportunity for reflection, an assessment tool and a tool for structuring quality and risk management. Any action taken is a result of the organisation's own process of assessment against the standards, not a bureaucratic consequence of the standards per se. In this process, a holistic view of the standards should be taken.

Second, the programme theory behind accreditation should be articulated and communicated clearly. It cannot be assumed to be apparent from the standards. This includes comprehensive communication of the view on standardisation.

Finally, attention must be taken to the design and format of specific accreditation standards. In the two published versions of the Danish hospital accreditation standards (IKAS, 2011, 2012), the measurable elements were organised into four steps to reflect the quality cycle (plan–do–study–act):

- Step 1: Guiding documents
- Step 2: Implementation of guiding documents
- Step 3: Quality surveillance
- Step 4: Quality improvement

This was devised for a didactic purpose but turned out to have unintended consequences. In the first version, requirements for quality surveillance were specified in the standards and were to a large extent of the process compliance type. In the second version, this specification of quality surveillance was largely removed, but since quality surveillance was still coupled to individual processes, it turned out to be difficult for hospitals to move to a more outcome-oriented and holistic approach to quality surveillance.

Taken together, the stepwise construction and the approach to quality surveillance have biased the use in the direction of 'the perverted quality cycle' and created the impression that QI begins with the writing of guidelines. This has been changed in special versions, including the one intended for use in the Faroese hospital system (IKAS, 2015). In versions for use by general practitioners and specialist physicians (IKAS, 2014, 2016), the need for written documentation has been heavily toned down.

CONCLUSION

We have reviewed how the Danish Healthcare Quality Programme has led to a proliferation of control requirements and, in general, a gap between, on the one hand, what is perceived as meaningful clinical quality assurance and improvement activity and, on the other hand, the requirements for control of compliance and documentation. In our opinion, if accreditation is done smartly it may offer a way of replacing ceaseless control with trust, based on an assessment of whether the system 'has what it takes'. Such an assessment will pay attention to professional judgement and variations in culture rather than rigid compliance with inflexible and static standards.

REFERENCES

Bogh SB, Falstie-Jensen AM, Bartels P et al. Accreditation and improvement in process quality of care: A nationwide study. *Int J Qual Health Care* 2015; 27: 336–343.

Braithwaite J, Greenfield D, Westbrook J et al. Health service accreditation as a predictor of clinical and organisational performance: A blinded, random, stratified study. *Qual Saf Health Care* 2010; 19: 14–21.

Brubakk K, Vist GE, Bukholm G et al. A systematic review of hospital accreditation: The challenges of measuring complex intervention effects. *BMC Health Serv Res* 2015; 15: 280. doi: 10.1186/s12913-015-0933-x.

Craig P, Dieppe P, Macintyre S et al. 2008. Developing and evaluating complex interventions: The new Medical Research Council guidance. *BMJ* 2008; 337: 979–983.

Dahler-Larsen P. *The Evaluation Society*. Stanford, CA: Stanford University Press, 2012.

Danish Health and Medicines Authority. Vejledning til læger, sygeplejersker, social- og sundhedsassistenter, sygehjælpere og kliniske diætister – Screening og behandling af patienter i ernæringsmæssig risiko, 2008 [in Danish]. http://sundhedsstyrelsen.dk/~/media/E47596E7CCB4491FB80FAB352750793C.ashx. Accessed 20 April, 2016.

Danish Ministry of Health. Nationalt kvalitetsprogram for sundhedsområdet 2015–2018, April 2015. Available at: http://www.sum.dk/Aktuelt/Publikationer/Nationalt-kvalitets-program-for-sundhedsomr-2015-2018-april-2015.aspx, see page 6 [in Danish]. Accessed 20 April, 2016.

Davidoff F, Dixon-Woods M, Leviton L et al. Demystifying theory and its use in improvement. *BMJ Qual Saf* 2015; 24: 228–238.

Dekker S. *Drift into Failure: From Hunting Broken Components to Understanding Complex Systems*. Farnham, England: Ashgate, 2011.

Dekker S. The bureaucratization of safety. *Safety Sci* 2014; 70: 348–357.

Dixon-Woods M, Baker R, Charles K et al. Culture and behaviour in the English National Health Service: Overview of lessons from a large multimethod study. *BMJ Qual Saf* 2014; 23: 106–115.

Dixon-Woods M, Bosk CL, Aveling EL, Goeschel CA, Pronovost PJ. Explaining Michigan: Developing an ex post theory of a quality improvement program. *Milbank Quarterly* 2011; 89: 167–205.

Falstie-Jensen AM, Larsson H, Hollnagel E et al. Compliance with hospital accreditation and patient mortality: A Danish nationwide population-based study. *Int J Qual Health Care* 2015a; 27: 165–174.

Falstie-Jensen AM, Nørgaard M, Hollnagel E et al. Is compliance with hospital accreditation associated with length of stay and acute readmission? A Danish nationwide population-based study. *Int J Qual Health Care* 2015b; 27(6): 20450–20457.

Fortune T, O'Connor E, Donaldson B. *Guidance on Designing Healthcare External Evaluation Programmes Including Accreditation.* ISQua, 2015, Dublin, Ireland. Available at: http://isqua.org/accreditation/reference-materials. Accessed 20 April, 2016.

Hasman A. 2012. IMIA accreditation of health informatics programs. *Yearb Med Inform* 2012; 7: 139–143.

Hinchcliff R, Greenfield D, Moldovan M et al. Narrative synthesis of health service accreditation literature. *BMJ Qual Saf* 2012; 21(12): 979–991.

Hollnagel E. *Safety-I and Safety-II: The Past and Future of Safety Management.* Farnham, England: Ashgate, 2014.

Holm-Petersen C, Wadmann S, Belén N, 2015. Styringsreview på hospitalsområdet: Forslag til procedure- og regelforenkling. [In Danish]. KORA 2015. http://www.kora.dk/udgivelser/udgivelse/i11098/Styringsreview-paa-hospitalsomraadet. Accessed 20 April, 2016.

IKAS. 2011. Accreditation standards for hospitals, 1st version, 2nd edn. 2011. http://www.ikas.dk/FTP/PDF/D11-6266.pdf. Accessed 20 April, 2016.

IKAS. 2012. Accreditation standards for hospitals, 2nd version, 2012. http://www.ikas.dk/FTP/PDF/D12-10072.pdf. Accessed 20 April, 2016.

IKAS. 2014. Accreditation standards for specialist physicians, 1st version. http://www.ikas.dk/FTP/PDF/D14-16135.pdf [In Danish]. Accessed 20 April, 2016.

IKAS. 2015a. Accreditation Standards for the Faeroese hospitals, 1st version, 2nd edn. http://www.ikas.dk/FTP/PDF/D15-9328.pdf [In Danish]. Accessed 20 April, 2016.

IKAS. 2015b. Stakeholder views on The Danish Healthcare Quality Programme – Accreditation standards for hospitals, 2nd version. http://www.ikas.dk/FTP/PDF/D15-2183.pdf [In Danish]. Accessed 20 April, 2016.

IKAS. 2016. Accreditation standards for general practice, 1st version. http://www.ikas.dk/FTP/PDF/D13-9227.pdf [In Danish]. Accessed 20 April, 2016.

ISQua. *Guidelines and Standards for External Evaluation Organisations*, 4th edn., Version 1.0. International Society for Quality in Health Care, Dublin, Ireland, 2013a.

ISQua. *Guidelines and Principles for the Development of Health and Social Care Standards*, 4th edn., Version 1.0. International Society for Quality in Health Care, Dublin, Ireland, 2013b.

Kondrup J, Johansen N, Plum LM et al. Incidence of nutritional risk and causes of inadequate nutritional care in hospitals. *J Clin Nutr* 2012; 21(6): 461–468.

Lie A, Bjørnstad O. Accreditation of occupational health services in Norway. *Occup Med (Lond)* 2015; 65(9): 722–724.

Mandag Morgen. Ny sundhedsmodel kan udløse revolution [New healthcare model may trigger revolution]. 28 January 2008. Available at https://www.mm.dk/ny-sundhedsmodel-kan-udl%C3%B8se-revolution. Accessed 20 April, 2016.

McMillen C, Zayas LE, Books S et al. Quality assurance and improvement practice in mental health agencies: Roles, activities, targets and contributions. *Adm Policy Ment Health* 2008; 35: 458–467.

Norén L, Ranerup A. Promoting competition in Swedish primary care. *J Health Organ Manag* 2015; 29(1): 25–38. doi: 10.1108/JHOM-04-2012-0080.

Paccioni A, Sicotte C, Champagne F. Accreditation: A cultural control strategy. *Int J Health Care Qual Assur* 2008; 21: 146–158.

Rasmussen LI. Every fifth hospital physician in the capital hospitals has protested against "documentation frenzy" [in Danish]. *Ugeskrift for Læger* Nov. 2011; 47(18): 2990–2991.

Roberts JS, Coale JG, Redman RR. A history of the joint commission on accreditation of hospitals. *JAMA* 1987; 258: 936–940.

Rozich JD, Howard RJ, Justeson JM et al. Standardization as a mechanism to improve safety in health care. *Joint Comm J Qual Patient Saf* 2004; 30: 5–14.

Schmaltz SP, Williams SC, Chassin MR et al. Hospital performance trends on national quality measures and the association with joint commission accreditation. *J Hosp Med* 2011; 6: 454–461.

Scrivens E. Recent developments in accreditation. *Int J Qual Health Care* Dec 1995; 7(4): 427–433.

Shaw C. External quality mechanisms for health care: Summary of the ExPeRT project on visitatie, accreditation, EFQM and ISO assessment in European Union countries. *Int J Qual Health Care* 2000; 12: 169–175.

Shaw CD, Braithwaite J, Moldovan M et al. Profiling health-care accreditation organizations: An international survey. *Int J Qual Health Care* 2013; 25: 222–231.

Suñol R, Vallejo P, Thompson A, Lombarts MJMH, Shaw CD, Klazinga N. 2009. Impact of quality strategies on hospital outputs. *Qual Saf Health Care* 2009; 18: i62–i68.

Verstraete A, Van Boeckel E, Thys M et al. Attitude of laboratory personnel towards accreditation. *Int J Health Care Qual Assur Inc Leadersh Health Serv* 1998; 11: 27–30.

Wears RL. Improvement and evaluation. *BMJ Qual Saf* 2015a; 24(2): 92–94. doi: 10.1136/bmjqs-2014-003889.

Wears RL. Standardisation and its discontents. *Cognition, Technology & Work* 2015; 17(1): 89–94.

Section III

Contemporary Nordic Research – Meso-Level Issues

9 A Multidisciplinary and Multifactor Approach to Falls Prevention

The RFPNetwork

Tarja Tervo-Heikkinen, Marja Äijö and Arja Holopainen

CONTENTS

INTRODUCTION

This chapter describes a project aimed at clarifying and unifying practices related to the prevention of falls in different contexts. Fall-related accidents are increasing especially among elderly people, and these accidents have both human and financial side effects (WHO, 2007; Stevens et al., 2006; Hartholt et al., 2012; Gill et al., 2013). Therefore, interventions were developed during this network project to prevent falls among the elderly living at home, in hospital or other contexts.

Falls and Fall-Related Accidents

Globally, falls and fall-related accidents are a serious health issue, especially among elderly people (WHO, 2007). Falls can cause both fatal (e.g. head traumas) and non-fatal (e.g. decreased functioning injuries) (Stevens et al., 2006; Korhonen, 2014). In addition, falls cause significant costs for healthcare. For example, falls increase hospitalisations and long-term nursing home admissions among elderly people after a hip fracture (Hartholt et al., 2012; Gill et al., 2013).

The age-standardised hip fracture rate has declined in western countries, but the incidence of hip fractures has increased numerically (Korhonen, 2014). When looking at gender differences, we found hip fractures to be more common among women than men, whereas some injuries, such as skull fractures, are more common among men (Orces, 2010; SotkaNET, 2015). In addition, there is significant variation in hip fracture rates among women from different nations (Litwic et al., 2012; Sosa et al., 2015). Among European women the highest fracture rates are seen in Scandinavia. One explanation can be vitamin D insufficiency, which is common in Scandinavian countries. It should be noted that incidence rates can vary widely even within the same country (Litwic et al., 2012). For example, some ethnic differences (e.g. bone material properties and bone mass) may partly explain the higher tendency for fractures in Norwegian women than in Spanish women (Sosa et al., 2015).

Falls, slips and trips are the most common reasons for accidental deaths in Finland. These accidents also cause more than 90% of hip fractures and more than half of brain injuries (Current Care Guidelines, 2008, 2011). According to Finnish statistics (SotkaNet, 2015), periods of care due to fall-related accidents are much more common among people aged 65 years and older than among younger people. The number of people over 65 years who sustain a hip fracture every year in Finland is almost 8000 (SotkaNet, 2015). Even though the incidence of hip fractures has declined in the new millennium in many western countries, including the Nordic countries (Kannus et al., 2006; Abrahamsen and Vestergaard, 2010; Stoen et al., 2012; Nilson et al., 2013), the absolute number of hip fractures has increased (Leslie et al., 2009; SotkaNet, 2015).

Research on the Subject in Brief

A lot of research on fall-related subjects has been conducted over the last decade globally (Joanna Briggs Institute, 2010; Cameron et al., 2012; Gillespie et al., 2012), as well as in Finland (Table 9.1). A common finding in all research is that fall-related

TABLE 9.1
Examples of Finnish Fall-Related Studies from 2007 to 2014

References	Setting, Material	Main Results
Hartikainen et al. (2007)	Systematic review (*N* = 29 studies) concerning medications as a risk factor for falls or fall-related fractures.	The main group of drugs associated with an increased risk of falling: psychotropics, antiepileptics and drugs that lower blood pressure.
Panula (2010)	Population-based study, *N* = 461, surgically treated hip fracture (1999–2000) patients. Incidence, morphometry, medication, mortality and cause of death were analysed.	Hip fractures: most commonly among women, occurred indoors and in institutions. Twenty-five per cent of patients had a previous fracture. Age-adjusted mortality was higher in men than in women. Fracture prevention, indoor safety measures and treatment of chronic lung diseases should be encouraged.
Karinkanta (2011)	RCT. *N* = 199, aged 70–78 years old women. Four training groups. The supervised training was three times a week for 12 months. Eighty-one per cent continued to the subsequent 1-year follow-up measurements (24-month assessment).	Twelve-month resistance and balance-jumping training, especially in combination, prevented functional decline by improving muscle performance and dynamic balance as well as self-rated physical functioning.
Piirtola (2011)	The incidence and predictors of fractures, functional decline and excess mortality due to fractures. *N* = 1177 (41% men) aged ≥ 65. Twelve years' follow-up.	Three hundred seven (26%) persons sustained altogether 425 fractures of which 77% were women. Reduced handgrip strength and BMI < 30 in women and a large number of depressive symptoms in men were independent predictors of fractures.
Salonoja (2011)	RCT multifactorial fall prevention study, 12 months, community-dwelling aged people 65 years and older, *N* = 591.	Twenty-five per cent used regularly psychotropic drugs, counselling and instructions decreased the regular use by 22% which had an effect to falls. The multifactorial prevention was successful in decreasing the incidence of falls in depressive or multiple fallers.
Korhonen (2014)	Register study. The epidemiology and trends in fall-induced deaths and injuries of older adults in Finland during 1970–2012 and predict the injury rates until the year 2030 by linear regression models.	The incidence of hospital-treated fall-induced injuries of older Finns rose from 1970 to 1990, but then the injury rates have declined. Incidence of severe head injuries and cervical spine injuries increased. The number of fall-induced injuries will increase during the coming decades.

accidents are caused by many factors. Therefore, their prevention also calls for different kinds of activities and multiprofessional actions in order to be effective.

Fall-Related Development Work and Education

It is important that professionals understand how they can prevent fall-related accidents; collaboration with educational organisations is therefore crucial. Theoretical background in this collaboration is integrative pedagogy which connects four elements of expertise. On an individual level, these elements are theoretical knowledge, practical expertise, self-regulation skills development and sociocultural information, which are manifested on a wider scale in the workplace (Tynjälä, 2008). This philosophy has previously been used in development work related to practical training situations in healthcare among healthcare students (Koskinen and Äijö, 2013; Äijö and Sirviö, 2012).

Integrative pedagogy integrates theory and practice. Theoretical knowledge includes information and facts about fall prevention, and the learning methods used are traditional, such as teaching using lessons and research articles. It is the responsibility of healthcare professionals to put this theoretical knowledge into action in their workplaces. Among students, integrative simulation situations at universities and vocational colleges and authentic learning experiences from practice placements or inter-professional falls prevention campaigns combine theoretical, practical and reflective elements of learning. Combining practical and theoretical information requires that students and professionals have self-regulation information. Self-regulation information includes a person's own learning habits, awareness of competence needs and one's own operational models with patients, for example, how to recognise and discuss risk of falling with patients in different kinds of situations and settings. Sociocultural knowledge consists of the unwritten rules of the work placement as well as the tools and equipment used. (Tynjälä, 2008.)

The purpose of this paper is to describe the development of the Regional Fall Prevention Network (RFPNetwork) and the actions implemented in it.

METHODS

Background for the Network

Interaction between people and organisations can be organised in three different ways: by forming hierarchies, markets and/or networks (Podolny and Page, 1998; Järvensivu et al., 2010). In a network, the distribution of tasks is resolved through flexible consultation and is based on trust. The network way to act is good when there is a need for innovative and flexible solutions (Järvensivu et al., 2010).

In Finland, the Ministry of Social Affairs and Health conducted a large healthcare project (2009–2011) called Attractive and Health Promoting Health Care. One part of the project was implementation of evidence-based nursing practices as well as development of different networks. This marked the beginning of fall prevention work in Kuopio University Hospital (KUH) and also the beginning of the collaborative project of KUH and the Nursing Research Foundation (Hotus) aimed at fall prevention in hospitals. One of the results of the project was the need to establish a regional network to promote fall prevention throughout the KUH region (Tervo-Heikkinen, 2011a,b). The need for the network arose from the Finnish national action plan for nursing management, where one of the aims was development of the management of nursing care by creating specific regional co-operation structures for healthcare practice, education and research (Ministry of Social Affairs and Health, 2009). The need for a multiprofessional network arose from the risks of falling and from the need of fall prevention, which requires multiprofessional work in order to succeed.

Development Work: How to Start the Operation of the Network

The Regional Fall Prevention Network (later RFPNetwork; AKE in Finnish) began its operation in May 2012 in the region of KUH district which covers, among others, the districts of North Savo, Central Finland and North Karelia, where the RFPNetwork operates in 2015. KUH is responsible for specialised medical care for almost one million people in Eastern and Central Finland. The RFPNetwork is a multiprofessional group with the aim of promoting prevention of falls and fall-related accidents in the whole area by supporting the introduction of evidence-based and consistent practices and guidelines. The main objective is that the number of fall accidents will not increase in the area. These goals are pursued through consistent and evidence-based instruction, good practices and education.

In Finland, specialised medical care hospitals are responsible for organising training courses for the staff in their region. Because of that obligation, the KUH organised a course in fall prevention before setting up the network. Afterwards, a leader for the network was appointed and organisations interested in the topic were identified. In the autumn of 2015, the network has grown to comprise 30 members from 11 organisations or groups of organisations.

Agenda for the First Meeting

Network co-operation adheres to the overall development cycle: planned activities, work, evaluation activities and acting as the basis for assessment (http://asq.org/learn-about-quality/project-planning-tools/overview/pdca-cycle.html). The fact that the establishment of the network has a legitimate reason does not mean that the network is efficient. It is essential for the operation of the network that it works flexibly and actively (Järvensivu and Möller, 2009; Järvensivu et al., 2010). Interaction between network members is based on genuine dialogue, based on the following principles: respect for others, listening to others, bringing up one's own voice and refraining from making hasty conclusions (Isaacs, 1999).

What are the tasks at the network's first meeting?

1. Familiarising the network members with each other.
2. Collecting all members' thoughts, expectations and wishes regarding the operation of the network.
3. Gathering information about practices that best serve the network's members and the organisations involved.

4. Gathering information about the measures or monitoring procedures that are used in the member organisations.
5. What expertise is required in addition to what is already included?
6. Agreeing on the issues that should be addressed at the next meeting.
7. Agreeing on meeting practices, such as frequency and length of meetings.

How did we do this in the RFPNetwork? The group that participated in the first meeting consisted of 12 representatives from five organisations. The spectrum of occupational groups has widened significantly after the first meeting. The objective has been to bring together professional groups whose activities can have the greatest influence on falls prevention.

At the first meeting, the convener made a presentation on falls and fall prevention for discussion by using research-based knowledge on the subject of falls and fall prevention. In brief, we had a uniform idea what the RFPNetwork should do and what its priorities were, as well as what further expertise we needed to complete the network. The RFPNetwork decided to meet four times a year. Between meetings, the network operates by e-mails and in small groups, if needed.

Network Meetings from the Second Meeting Onwards

Once all basic practicalities had been settled, it was time to focus on action:

1. The group had been supplemented with new expert members.
2. Any new organisations interested in the operation of the network could join in.
3. Formulated the network action plan: ideas, expectations and wishes formed the basis for an action plan.
4. Prioritisation of tasks.
5. The network had to be open to new ideas and tasks.
6. Agreement that all members of the network had to do their part.

As described earlier, the RFPNetwork recruited new expert members from old and new organisations. Although it is important to complement the expertise of the network, it is not optimal for the network to grow too much or too rapidly. The network approach means making decisions together with shared understanding. In a fast-growing network, a lot of time has to be spent on grouping the members; and as a result, decisions do not get made (Järvensivu et al., 2010).

Operational Area of the Regional Fall Prevention Network

The goal for the network was to have an impact in the whole region. In the autumn of 2015, there were 30 members from 11 organisations or organisation groups in the RFPNetwork. They represented different kinds of hospitals, health centres, nursing homes, homecare or educational organisations in the area. Network members also represented many professions, such as physicians, geriatricians, nutrition therapists, lecturers, researchers, physiotherapists, patient safety managers, registered nurses, osteoporosis nurses, nursing managers, pharmacists, clinical nurse specialists, emergency field directors and development managers. Some of the members joined the network because of their own interest in the subject and the desire to improve patient safety. Others had been invited to join the network because they have valuable expertise on the topic. The important thing was to identify the network's needs and the benefits of multidisciplinary actions (Järvensivu and Möller, 2009). In autumn 2015, the members represented a geographically large area, as can be seen in Figure 9.1.

In the RFPNetwork, this means that all member organisations use the same consistent falls prevention practices, procedures and guidelines and that the educational organisations teach these to the students, for example to nurses, physiotherapists, practical nurses and medical students. The RFPNetwork is an example on how different professionals with a background from education and clinical practices can develop practices together. In addition, students from different educational organisations and professions are key actors and developers in this kind of network together with healthcare professionals and teachers.

Hotus – Preventing Fall-Related Accidents through Knowledge Sharing and Collaboration

The goal of the Nursing Research Foundation (Hotus) is to disseminate evidence-based guidelines and information about evidence-based practices to healthcare professionals. A good example is the falls prevention project in the KUH region where Hotus and the RFPNetwork have been co-operating partners.

Hotus is one of the partners involved in disseminating this network model nationally. With the goal of applying best practices at the regional level, beyond individual hospitals or healthcare centres, this project has brought together a variety of healthcare and educational organisations in the RFPNetwork.

RESULTS

In the RFPNetwork, ideas and expectations of all members are brought together. These ideas and expectations include consistent education and training, consistent instructions and recommendations, guidelines and educational materials, monitoring of accidental falls from different registers and, later, increasing visibility of the RFPNetwork. The network made an action plan for 2 years at a time. The first thing was to draw up consistent good practice recommendations that are suitable to all organisations and units. It was clear that the recommendations would have to be evidence based. The next task was to find out what kind of instructions about fall prevention were available to the general public and what might be needed. As a result, we started to draw up the 'Stay Up!' guide, published in Finnish, Swedish and English. In the same way, we also produce different types of material for practice.

Systematic Follow-Up of Falls

The work of the network has to be monitored. The topic of the RFPNetwork is very multidimensional, and monitoring the impact of its operations is challenging.

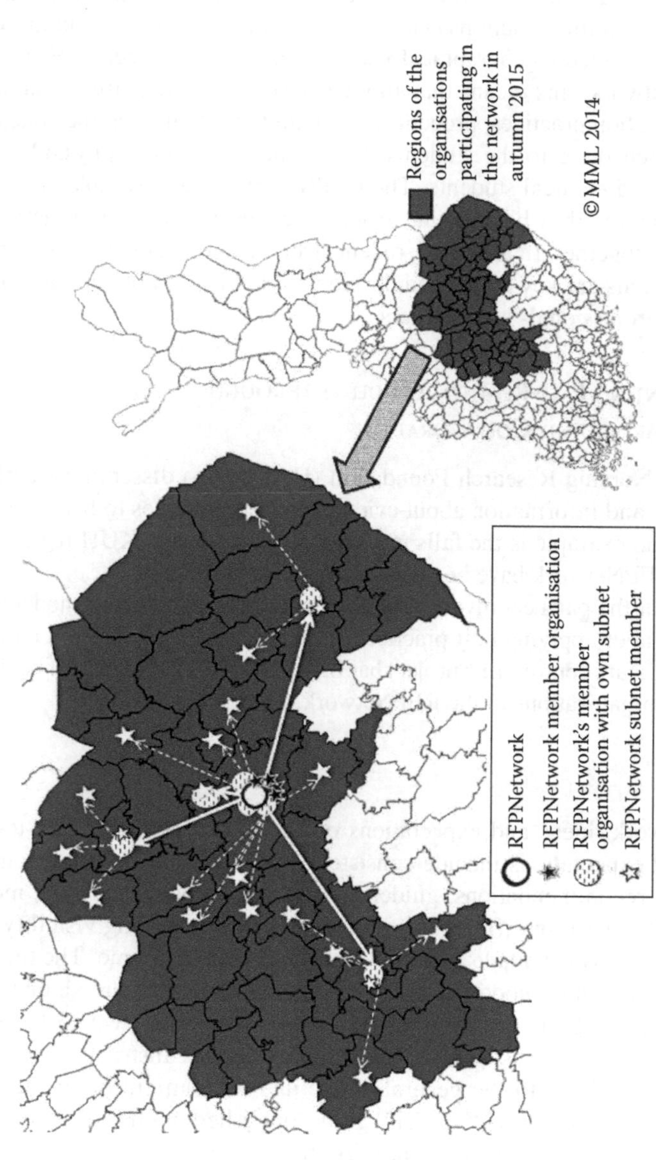

FIGURE 9.1 The RFPN operation area in 2015, the participating hospital regions in Finland and RFPNetwork member organisations and subnetworks with their members.

However, at the onset of work, the RFPNetwork agreed on a common follow-up of fall-related accidents on organisational and national level. It was decided that organisational level follow-up would be performed every year from the Reporting System for Safety Incidents in Health Care Organizations (HaiPro) (http://awanic.com/haipro/eng/). The organisations that do not use HaiPro record accidents in accordance with their own registers. The visibility of the network will also be monitored by surveys.

Regional Process Description of Fall Prevention

One of the tasks for the RFPNetwork was the description of the regional fall prevention process. The process of description begins in the home, that is the place with the first opportunity to have a preventive effect on falls prevention. The process goes through falls prevention planning in different organisations, from homes to hospital and back. It includes all RFPNetwork tools to measure and prevent the risk of falling and instructions on what to do if someone actually does have a fall.

All completed materials have been included in the process description, which will be supplemented as and when new material is completed. The material is distributed electronically to members of the network. We have also had an opportunity to share open materials via the Hotus homepages (http://www.hotus.fi) and, as of May 2015, the KUH website (http://www.psshp.fi/ake in Finnish and in English pages http://www.psshp.fi/rfpnetwork)

Implementation of the Falls Prevention Process

Once the basics for the fall prevention process and network are in place, the process has to be implemented into practice. It is important to understand that the basic network cannot function and grow endlessly. That is why this kind of network needs subnetworks. Subnetworks can operate within an organisation which, for example, brings together different units at a hospital or health centre so that implementation is made possible. Every unit needs a key person in order to ensure implementation. The key person is familiar with the phenomenon, instruments and materials, so she/he is able to guide colleagues.

The advantages of the regional network to implementation are

1. Uniform processes
2. Uniform instructions and checklists
3. A uniform and systematic follow-up system, for example concerning falls in the area
4. The use of uniform tools, such as fall risk assessment tools
5. Uniform guidebooks to the general public

Forming Subnetworks

As noted earlier, it is important to complement the expertise of the network, but it is not optimal for the network to grow too much or too rapidly. The network needs different types of expertise, but the implementation of consistent practices and

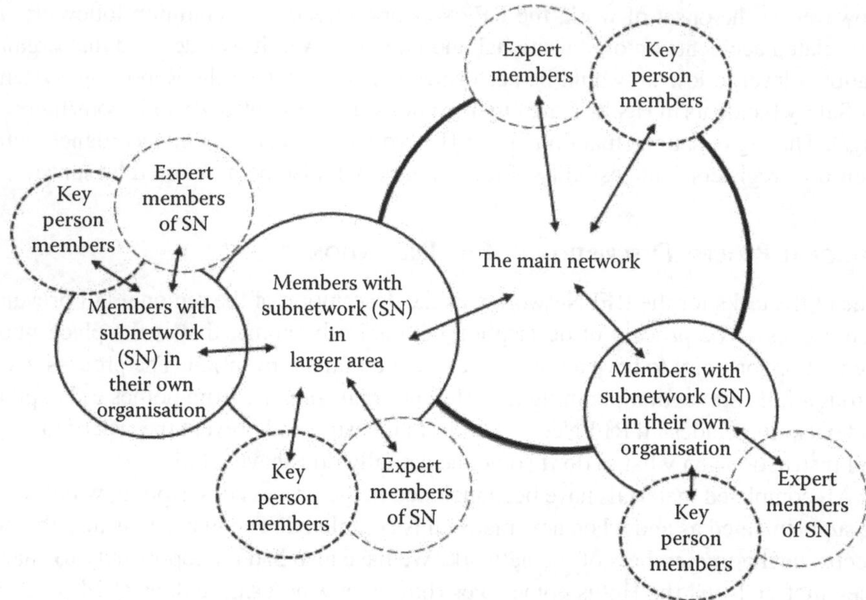

FIGURE 9.2 Forming a large network with subnetworks.

guidelines also requires more than the input from the regional network. There has to be actors who are closer to practice than the main network.

The RFPNetwork organisations have therefore set up subnetworks. A subnetwork can operate within one organisation or, more widely, for example, within a municipality or an even larger district. For example, Hoivakoti Aurinkopuisto is a private nursing home with a subnetwork on one organisation level. Siilinjärvi, on the other hand, is a municipality and has a subnetwork that brings together different parts of social and healthcare services. The establishment of a subnetwork to the RFPNetwork has been initiated at the Central Finland Health Care District, comprising 21 municipalities (Figure 9.2).

If all these members belonged to a single network it would suffocate the operation of the network. Subnetworks and practical actors, which we call key persons, are also needed for implementation. It must, however, be ensured that the voice of the key persons is heard in the main network as well.

Collaboration in the Regional Fall Prevention Network – How Do the Subnetworks Operate?

The municipality of Siilinjärvi has been involved in the RFPNetwork from the beginning and their subnetwork actions are promising. Very early on, they formed a subnetwork of their own. It consists of representatives from different areas of social and healthcare and various professions. This subnetwork has a fixed meeting schedule, and they are working actively to bring RFPNetwork's consistent practices and materials to the workplace.

The Siilinjärvi subnetwork has drawn up a plan for the years 2015 and 2016. One of the tasks of the RFPNetwork is to increase the visibility of the network and to raise people's awareness of their own opportunities to exert influence on falls prevention. Information directed at the general public has also been added to the municipality's homepage; they have organised an 'Elderly week' on the topic of falls prevention and worked together with different actors to arrange a series of lectures entitled 'Your Health'.

Educational Development Work

Falls prevention work has strengthened the collaboration between educational organisations and workplaces. This collaboration enables students' learning through a variety of development tasks, theses and research. Students have carried out falls prevention campaigns and, at the same time, learned about project planning, implementation and evaluation. The project design also promotes the development of project management skills, increases theoretical knowledge of falls prevention and supports conversation skills with customers or patients. Data collected from different campaigns of falls prevention and development work give students material for their theses.

The consistent practices, procedures and instructions developed by the RFPNetwork form key contents for teaching. These good practices are taught and practised, for example, in simulated situations at educational organisations. Familiarity with practices and forms, for example Falls Risk for Older People (FROP-Com Screen, Russell et al., 2008) or RFPNetwork's checklist for fall prevention, and training in their use, contributes to students' skills to work with patients and clients. In addition, knowledge of falls prevention activities and prevention process contributes to correct guidance of patients and the use of forms during the students' clinical practice periods. This has increased students' knowledge and skills to work with patients and as part of multidisciplinary teams during training periods.

Teaching simulation situations can be highlighted, and they contribute to the development of falls prevention work. For example, polypharmacy, sedating medications and orthostatic hypotension have been identified as risk factors among people who have fallen at home or in hospital (Ambrose et al., 2015). When significant risk factors are recognised, healthcare professionals are able to use preventive methods to reduce the risks. This increases readiness to detect the risks and prevent falls. Issues relating to the safety of elderly patients' treatment and rehabilitation also comprise the emphasis on teaching, for example, advice and guidance relating to safe movement in patients' homes as well as in hospitals and nursing homes. The Stay Up! Guide has already been in use in teaching situations, among others. Development work focusing on good practices and teaching increases students' expertise and professional skills to implement fall prevention measures, particularly among the elderly. In addition, healthcare professionals need to update their knowledge and skills as well. Lifelong learning is key in carrying out and supporting fall-related preventive work among healthcare professionals and local residents in the whole area.

DISCUSSION

The current period of the project will end in December 2015, and the RFPNetwork's task for the rest of the year is to draw up an action plan for the next 2 years. Ideas and needs have already been submitted to the network from subnetworks, expert members and key persons. Regular monitoring and follow-up will continue. New statements are already on the stocks, and new ideas will continue to be received. It is also important to highlight the visibility of the network's actions by writing articles, through participation in conferences, and by working actively with different organisations, such as universities. University students in different professions and at different levels can write theses on topics related to falls prevention and fall-related issues.

In the future, promotion and research is one part of the collaboration between the network organisations. This also promotes follow-up and visibility of the network. Implementation data will be collected from healthcare professionals and people living in the operation area. In this project, we used falls as an example in simulation and integrative pedagogy. One of the topics for future research is healthcare students and their learning experiences from simulation and integrative pedagogy.

Success Factors for Multiprofessional RFPNetwork

1. Multidisciplinary experts are involved – commonly agreed objectives and measures.
2. Network operation organised by an influential organisation.
3. Network convened by experts on the topic.
4. The participants and organisations involved are committed to the network.
5. The network is supported by the management of the organisations involved.
6. Management ensures the availability of network's resources.
7. The network has uniform aims and has planned how to monitor progress (aims and outcomes).
8. In order to be successful, implementation calls for
   a. Key personnel at the units.
   b. Consistent and regular education.
   c. Uniform good practices suitable for different areas in the healthcare context (measurement, monitoring and reporting).
   d. Uniform education materials for the general public and staff.
9. Visible and open sharing within the network.
10. Multiprofessional, encouraging and supportive co-operation.

REFERENCES

Abrahamsen, B. and Vestergaard, P., 2010. Declining incidence of hip fractures and the extent of use of anti-osteoporotic therapy in Denmark 1997–2006. *Osteoporosis International* 21(3), 373–380.

Äijö, M. and Sirviö, K., 2012. An integrative learning model for a mobile unit. In: K. Sirviö and M. Äijö (eds.), *Suupirssi – Expertise on the Move*. Kuopio, Finland: Savonia University of Applied Science, pp. 45–48.

Ambrose, A.F., Cruz, L. and Paul, G., 2015. Falls and fractures: A systematic approach to screening and prevention. *Maturitas* 82(1), 85–93.

Cameron, I.D., Gillespie, L.D., Robertson, M.C., Murray, G.R., Hill, K.D., Cumming, R.G. and Kerse, N., 2012. Interventions for preventing falls in older people in care facilities and hospitals. *Cochrane Database of Systematic Reviews* [online], (12), CD005465. Available at: http://onlinelibrary.wiley.com/doi/10.1002/14651858.CD005465.pub3/abstract (accessed 16 October 2015).

Current Care Guidelines, 2008. Aivovammat [Brain injuries] (in Finnish). [Online]. Available at: http://www.terveysportti.fi/xmedia/hoi/hoi18020.pdf (accessed 16 October 2015).

Current Care Guidelines, 2011. Lonkkamurtuma [Hip fracture] (in Finnish). [Online]. Available at: http://www.terveysportti.fi/xmedia/hoi/hoi50040.pdf (accessed 16 October 2015).

Gill, T.M., Murphy, T.E., Gahbauer, E.A. and Allore, H.G., 2013. Association of injurious falls with disability outcomes and nursing home admissions in community-living older persons. *American Journal of Epidemiology* 178(3), 418–415.

Gillespie, L.D., Robertson, M.C., Gillespie, W.J., Sherrington, C., Gates, S., Clemson, L.M. and Lamb, S.E., 2012. Interventions for preventing falls in older people living in the community. *Cochrane Database of Systematic Reviews* [online], (9), CD007146. Available at: http://onlinelibrary.wiley.com/doi/10.1002/14651858.CD007146.pub3/abstract (accessed 16 October 2015).

Hartholt, K.A., Polinder, S., Van der Cammen, T.J., Panneman, M.J., Van der Velde, N., Van Lieshout, E.M., Patka, P. and Van Beeck, E.F., 2012. Costs of falls in an ageing population: A nationwide study from the Netherlands (2007–2009). *Injury* 43(7), 1199–1203.

Hartikainen, S., Lönnroos, E. and Louhivuori, K., 2007. Medication as a risk factor for falls: Critical systematic review. *Journal of Gerontology: Medical Sciences* 62A(10), 1172–1181.

Isaacs, W., 1999. *Dialogue: The Art of Thinking Together.* London, U.K.: Bantam Doubleday Dell Publishing Group.

Joanna Briggs Institute, 2010. Interventions to reduce the incidence of falls in older adult patients in acute care hospitals. *Best Practice: Evidence-Based Information Sheets for Health Professionals* 14(15), 1–4.

Järvensivu, T. and Möller, K., 2009. Metatheory of network management: A contingency perspective. *Industrial Marketing Management* 38, 654–661.

Järvensivu, T., Nykänen, K. and Rajala, R., 2010. Verkostojohtamisen opas: Verkostotyöskentely sosiaali- ja terveysalalla [The Network Management Guide: Working with a network of social and health care] (in Finnish). [Online]. Available at: http://verkostojohtaminen.fi/wp-content/uploads/2011/01/VerkostojohtamisenOpas.pdf (accessed 16 October 2015).

Kannus, P., Niemi, S., Parkkari, J., Palvanen, M., Vuori, I. and Järvinen, M., 2006. Nationwide decline in incidence of hip fracture. *Journal of Bone and Mineral Research* 21(12), 1836–1838.

Karinkanta, S., 2011. To keep fit and function – Effects of three exercise programs on multiple risk factors for falls and fractures in home-dwelling older women. Academic dissertation, University of Tampere, Tampere, Finland.

Korhonen, N., 2014. Fall-induced injuries and deaths among older Finns between 1970 and 2012. Academic dissertation, University of Tampere, Tampere, Finland.

Koskinen, L. and Äijö, M., 2013. Development of an integrative practice placement model for students in health care. *Nurse Education in Practice* 13(5), 442–448.

Leslie, W.D., O'Donnell, S., Jean, S., Lagace, C., Walsh, P., Bancej, C., Morin, S., Hanley, D.A., Papaioannou, A. and Osteoporosis Surveillance Expert Working Group, 2009. Trends in hip fracture rates in Canada. *JAMA* 302(8), 883–889.

Litwic, A., Edwards, M., Cooper, C. and Dennison, E., 2012. Geographic differences in fractures among women. *Womens Health* 8(6), 673–684.

Ministry of Social Affairs and Health, 2009. Johtamisella vaikuttavuutta ja vetovoimaa hoitotyöhön [An action plan for the years 2009–2011. Increasing the effectiveness and attraction of nursing care by means of management] (Abstract in English). [Online]. Available at: http://urn.fi/URN:ISBN:978-952-00-2919-7 (accessed 16 October 2015).

Nilson, F., Moniruzzaman, S., Gustavsson, J. and Andersson, R., 2013. Trends in hip fracture incidence rates among the elderly in Sweden 1987–2009. *Journal of Public Health (Oxford)* 35(1), 125–131.

Orces, C.H., 2010. Trends in hospitalization for fall-related injury among older adults in the United States, 1988–2005. *Ageing Research* 1(e1), 1–4.

Panula, J., 2010. Surgically treated hip fracture in older people. With special emphasis on mortality analysis. Academic dissertation, University of Turku, Turku, Finland.

Piirtola, M., 2011. Fractures in older people – Incidence, predictors and consequences. Academic dissertation, University of Turku, Turku, Finland.

Podolny, J.M. and Page, K.L., 1998. Network forms of organization. *Annual Review of Sociology* 24(1), 57–76.

Russell, M.A., Hill, K.D., Blackberry, I., Day, L.M. and Dharmage, S.C., 2008. The reliability and predictive accuracy of the falls risk for older people in the community assessment FROP-Com Screen (Falls Risk for Older People) tool. *Age and Ageing* 37(6), 634–639.

Salonoja, M., 2011. Fall-risk-increasing drugs: Multifactorial fall prevention among the aged in Pori. Academic dissertation, University of Turku, Turku, Finland.

Sosa, D.D., Vilaplana, L., Güerri, R., Nogués, X., Wang-Fagerland, M., Diez-Perez, A. and Eriksen, E.F., 2015. Are the high hip fracture rates among Norwegian women explained by impaired bone material properties? *Journal of Bone and Mineral Research* 30(10), 1784–1789.

SotkaNet, 2015. Statistical information on welfare and health in Finland. National Institute for Health and Welfare. Available at: https://www.sotkanet.fi/sotkanet/en/taulukko/?indicator=s06MsPaPiDcEAA==®ion=s07MBAA=&year=sy4rsTbS0zUEAA==&gender=m;f;t&abs=f&color=f_(accessed 16 October 2015).

Stevens, J.A., Corso, P.S., Finkelstein, E.A. and Miller, T.R., 2006. The costs of fatal and non-fatal falls among older adults. *Injury Prevention* 12(5), 290–295.

Stoen, R.O., Nordsletten, L., Meyer, H.E., Frihagen, J.F., Falch, J.A. and Lofthus, C.M., 2012. Hip fracture incidence is decreasing in the high incidence area of Oslo, Norway. *Osteoporosis International* 23(10), 2527–2534.

Tervo-Heikkinen, T., 2011a. Terveydenhuollon organisaatioiden verkostoitumisen kuvaus. Yhteistyöverkostojen toiminnan kuvaaminen [Description of the health care organizations networking. Description of Co-operation Networks] (in Finnish). [Online]. Available at: http://www.vete.fi (accessed 16 October 2015).

Tervo-Heikkinen, T., 2011b. Yhtenäisten näyttöön perustuvien käytäntöjen kehittäminen. Kaatumisten ehkäisy KYSissa pilotoinnin käytännön etenemisen kuvaus [Development of evidence-based practices. Fall prevention in KUH, the pilot study description] (in Finnish). [Online]. Available at: http://www.vete.fi (accessed 16 October 2015).

Tynjälä, P., 2008. Perspective into learning at the workplace. *Educational Research Review* 3(2), 130–154.

World Health Organization (WHO), 2007. WHO global report on falls prevention in older age. [Online]. Available at: http://www.who.int/violence_injury_prevention/publications/other_injury/falls_prevention.pdf?ua=1 (accessed 16 October 2015).

10 Coordination of Discharge Practices for Elderly Patients in Light of a Norwegian Healthcare Reform

Heidi Helen Nedreskår and Marianne Storm

CONTENTS

INTRODUCTION

Hospitals and municipalities expect their healthcare personnel to conform to different values and professional traditions. Hospitals focus on treatment, whereas municipalities focus on care, function and coping. These different tasks are performed separately, rather than in collaboration, even when such collaboration would be the best solution for the patient (Danielsen and Fjær 2010; Glouberman and Mintzberg 2001). Many studies have revealed the lack of coordination when discharging older patients from hospital to home as a significant problem (Aase et al. 2013; Dahl et al. 2014; Laugaland et al. 2011, 2012). Hospital discharge is a critical phase in patient recovery, and shorter hospital stays and rapid discharge require coordination between health personnel across organisations to ensure adequacy of services and to avoid readmissions (Storm et al. 2014). Elderly patients with both physical and cognitive health challenges represent a significant proportion of health service users in both hospitals and municipalities. Because Norway has a twofold healthcare system and because the two services have different legally mandated tasks, it will be these elderly patients who are most often moved between hospital and municipality health services (Tjora and Tøndel 2012). Coordination of health services is essential for these patients to receive proper treatment and care.

Øgar and Hovland (2004) defined coordination as 'the information exchange, knowledge transfer, and assignment of responsibilities and duties to safeguard patients' needs, and the overall health policy objectives and regulatory requirements that apply to health services' (p. 166). Haggerty et al. (2003) have studied the concept of continuity in healthcare services. Continuity of care is a multidimensional concept closely related to coordination and comprising the following dimensions: informational continuity, management continuity and relational continuity. 'The importance attached to each type differs according to the providers and the context of care, and each can be viewed from either a person focused or disease focused perspective' (Haggerty et al. 2003, p. 1220). Better coordination follows from continuity (Haggerty et al. 2003).

Management continuity is a structural dimension characterised by political healthcare guidelines and financial arrangements for the patient's daily care (Hellesø 2012). According to Haggerty (2003), management continuity is achieved when multiple services are delivered in a complementary and timely manner. Sharing of management plans or care protocols facilitates management continuity and helps both patients and healthcare personnel to secure future care. An important aspect of management continuity is flexibility in adapting care to changes in patients' needs (Haggerty et al. 2003).

Continuity of information means that the healthcare personnel responsible for a given patient are also in possession of adequate information about that patient. Additionally, the patient should possess adequate information about treatment, prognosis and future plans for both specialist health services and the municipality's care services (Hellesø 2012). Haggerty (2003) described informational continuity as 'the use of information on past events and personal circumstances to make current care appropriate for each individual' (p. 1220).

For a healthcare system to legitimately claim continuity of service, the patient must experience care as integrated, connected and coherent in relation to his/her medical and personal needs (Haggerty et al. 2003). Relational continuity implies 'an ongoing therapeutic relationship between a patient and one or more providers' (Haggerty et al. 2003). It means that healthcare personnel are familiar with the patient and his/her medical history, that all parties agree on the plan for care and that the patient knows about prospective services. 'The experience of continuity may differ for the patient and the providers, posing a challenge to evaluators' (Haggerty et al. 2003, p. 1221).

The purpose of this study is to investigate how healthcare and administrative personnel perceive coordination of health services between hospital and municipality in two healthcare trusts and their respective municipalities in Norway, as well as to explore how personnel coordinate healthcare services at discharge from hospital to short-term municipal placements. The study focuses on discharge to ordinary, short-term placements and to a transitional ward established with the purpose of receiving patients ready for discharge.

The following research questions will be answered:

1. How does coordination between hospital and municipality take place when implementing new discharge agreements?
2. How do healthcare personnel and administrative personnel characterise coordination of discharge practices for elderly patients?

METHODOLOGY

A qualitative study was conducted using individual interviews (Kvale and Brinkmann 2009) of a strategic selection of healthcare personnel and administrative employees at a university hospital and a county hospital, as well as with personnel in some of the respective municipalities. The aim was to understand informants' subjective experiences of events in relation to discharging patients from the hospital and transferring them to short-term placements or transitional ward at a nursing home.

Informants' employers asked employees whether they wished to participate in an interview, and interviews were conducted at the workplace. Before the interviews, informants received information about the study and notification that participation was voluntary (Kvale and Brinkmann 2009; Thagaard 2010). Informants were also asked to describe their experiences of the flow of information from hospital to municipality upon patient discharge to short-term facilities. Interviews were tape-recorded and transcribed verbatim.

Five of the interviews were conducted in October 2012 and 11 of the interviews in the spring of 2013. The interviews were conducted in two different counties, in two small rural municipalities affiliated to a county hospital and in a large municipality affiliated to a university hospital (Table 10.1).

A qualitative content/text analysis was performed (Graneheim and Lundman 2004). After reviewing the tape recordings and the transcribed interviews, the data were systematised and analysed, resulting in four themes. Both authors participated in the analysis.

ETHICS

The study is part of the research project 'Quality and safety' (Aase et al. 2013). The project was certified by the Regional Committees for Medicine and Health Research Ethics on October 19, 2011 (ref. no. 1978). All informants provided written consent to participate in the study.

TABLE 10.1
Informants' Age, Gender, Place of Work and Job Title at the Time of the Interviews

Study Informants	Municipalities	Hospitals
Age	38–62 years	40–61 years
Gender	8 women	7 women, 1 man
Educational background	Bachelor of nursing, bachelor of physiotherapy specialist training in geriatrics, palliative care and management and master degree	Bachelor of nursing, master of social science, master degree and specialist training in management
Work experience	13–34 years	14–33 years
Workplace	Nursing homes, Patient Service Unit	Surgical and medical wards, coordination section
Positions	Ward leader, head of nursing home, patient coordinator, secretary for receipt of request from hospital	Employees and leaders at coordination section, ward leader, administrative nursing leader

Context

This study was conducted in two counties in Norway, located within the same regional health authority. The urban hospital is affiliated to 18 municipalities, and the county hospital is affiliated to 21 municipalities. Many of Norway's municipalities have small populations (Romøren et al. 2011) and are geographically far from a hospital.

The Norwegian healthcare system is divided into two separate organisational levels: primary care and specialised secondary. Local municipalities are responsible for primary care including nursing homes, home care, public health nursing, physiotherapy, ergo therapy and general practice. Specialised healthcare is provided in state-owned hospitals and organised by four regional health authorities. These two levels act according to different laws, regulations, goals and tasks (Laugaland 2015).

The introduction of the Coordination Reform in 2012 required of hospitals and municipalities a greater degree of formalised cooperation (Law of Primary Health and Care Services 2011; Law of Specialist Care 2009; Report to Parliament [No. 47] 2008–2009). The intention of the reform is to ensure that healthcare services improve continuity and coordinated services (Report to Parliament [No. 47] 2008–2009).

The Coordination Reform and the Coordination Agreement (Samhandlingsavtaler) require hospitals and municipalities to provide an explicit address for inquiries about patients and service users (Coordination Agreement 2012 and Report to Parliament [No 47] 2008–2009). The term 'Coordinating Unit' is used by both municipal and specialist services. The Coordinating Unit responds to inquiries, provides administration and secures continuity of service provision between the primary and specialist healthcare (Det kongelige Helse- og Omsorgsdepartementet Meld. St. 16 (2011–2015), Helsedirektoratet 2012). The Coordination Reform stipulates a preference for electronic communication between hospitals and municipalities when exchanging health information regarding patients (Report to Parliament [No. 47] 2008–2009). Electronic message exchange (EME) among hospitals, municipalities (e.g. home care, nursing homes, Patient Service Units) and general practitioners is in an initial phase, and there are major differences among practitioners in their use of the new technology. Electronic communication will be fully developed in all hospitals and municipalities in the coming years (Bergmo et al. 2013).

The Coordination Reform and related laws require specialist health services and municipalities to form binding agreements in order to contribute to cooperation and coordination between levels of administration and care, ensuring better continuity of care and coordinated services (Det kongelige helse- og Omsorgsdepartementet Meld. St. 16 (2011–2015); Law of Primary Health and Care Services 2011; Law of Specialist Care 2009; Report to Parliament [No. 47] 2008–2009). Sub-agreements and one overarching agreement have been produced according to the law in all municipalities and for specialist services in Norway. Agreements are binding for both parties and describe the tasks appropriate to the respective organisations. Breach of agreements results in a written notice of deviation and, possibly, sanctions.

This study is concerned with one sub-agreement, which deals with allocation of tasks/duties and responsibilities regarding somatic patient stays in and discharges from special health services to municipal health services. According to this sub-agreement, the municipality is financially responsible for patients ready to discharge. If a municipality does not immediately receive patients ready for discharge, it must pay a penalty fee to the hospital. Should the patient receive services from the municipality, the hospital must notify the municipality about admittance within 24 hours (early notification). According to the sub-agreement, written notice must also be provided in cases where patients have completed treatment before discharge. The ready-to-discharge notice designates a patient's completion of treatment, and written notice is sent from the hospital to the municipality.

RESULTS

The results are presented according to the four themes revealed through data analysis: implementing new agreements, organisation and information flow, when is a patient ready for discharge from hospital? and competence and experience in health personnel.

Implementation of New Agreements

The introduction of the agreements of the Coordination Reform in 2012 required comprehensive training of employees in hospitals and municipalities. For effective coordination, the employees of both organisations need adequate knowledge of the agreements and of how to put them into practice when transferring patients.

Informants from the hospitals and in the municipality experienced that the work related to the agreements is anchored by management and is executed by managers at different levels of the organisation. The work is described as encompassing and time-consuming. Informants at both hospitals and in the municipality explained that a range of different methods are used to develop and implement the agreement.

Findings show that there are several similarities between the urban and the rural approach to implement the agreements. Both have engaged administrative and healthcare personnel in working groups together with the management to come to an agreement for both municipalities and hospitals. Management in both hospitals and urban and rural municipalities was concerned with providing information to employees during the process of implementing the agreements. The Internet and newsletters were used to gather and disseminate information. The rural hospital and municipalities had a specific project group focusing on the discharge agreement, because they expected this agreement to represent a difficult task for both organisations. They arranged a meeting with personnel from both the hospital and the municipalities where personnel completed a questionnaire about their experiences of the discharge agreement. The county hospital and the rural municipalities also had user representative in the group implementing the discharge agreement.

The urban hospitals have established coordination contacts (nurses with extra knowledge of the functions and services in the municipalities) to improve coordination with the municipalities. To improve coordination, the urban hospital continued with and strengthened the mutual practical consultant arrangements they already had in place with municipal doctors and nurses.

Informants from the Coordination section at the urban hospital related that they previously had agreements about patients ready for hospital discharge with many local authorities, but that now these agreements are legally mandated. Informants stated that the employees have been involved in coordination work in varying ways with virtually all sub-agreements, just not with the overarching agreement. They described the agreement on discharge-ready patients as a detailed distribution of work for service providers at both levels. Informants perceived the concrete nature of the agreement as an advantage. The agreement, they said, is easier to follow through and practice when smaller matters are left to the judgement of healthcare personnel. The informants reported that their organisations had given significant attention to new routines for notifying municipalities of patients who have completed treatment. They said that healthcare personnel are drilled in the agreement-specified procedures related to patient discharge from hospital.

The informants from the county hospital mentioned that, previously, coordination between the two service levels had been characterised by many small, uncoordinated municipalities and large hospital, which gave the hospitals power. The hospitals are affiliated to many municipalities and, prior to the agreement, the municipalities had each negotiated separately with the hospital, leaving them with less power than the hospitals. One participant was of the opinion that this had changed for the better under the coordination agreement:

> The coordination environment changed ... This is the idea of the Coordination Reform, to give greater equitability. And it was a very good process, and I found that we were fortunate to have the agreement and a sense of equality emerged.
>
> **Coordination section, county hospital**

Many of the informants said that when the municipalities stand together, they become a greater power. Although some municipalities are quite small, coordination gives the municipalities a strengthened position, which also reflects in specific, tangible cooperation at the departmental level.

According to the informants, there is a greater focus on coordination in the organisations, and the sub-agreement leads to earlier and more frequent contact between hospital and municipality on discharge of patients. Municipalities now have few or just one telephone number, which the hospitals use when planning for discharge of patients, and there are prepared checklists and standard forms for contact between organisations.

Organisation and Information Flow

According to the informants, Coordinating Units are organised in different ways among municipalities. In the urban municipality, the Patient Service Unit functions as the Coordinating Unit. There are four units serving the entire municipality. Patients seeking healthcare services are referred to the Patient Service Unit, where each is assigned a permanent patient coordinator. The Health and Social Service head is the top manager and professional responsible for the decisions taken. One of the informants at the Patient Service Unit enlarged on the role:

> First and foremost it is to coordinate services to users … to see that they get the appropriate help. We enter into dialogue with next of kin, have home visits to patients, see how it goes and think further. With changes in need the idea is to see the whole, make a few summaries, and be available.
>
> **Patient Service Unit, urban municipality**

The strength here, according to this informant, lies in the fact that one gains an early oversight of the patient's needs and can make a plan in conjunction with other help agencies and with next of kin. Informants from the two rural municipalities reported that they also used the term 'Coordinating Unit', but the units were differently organised and the daily activity functioned differently. Both had telephone numbers that hospital personnel could call. The person who answered the phone could vary, they reported, and the call could then be passed on to the duty telephone at the nursing home or directly to the home care nurse. In both municipalities, the ward leaders at nursing homes played a key role in servicing the contact phone line, even though calls often concerned patients that were not being discharged to the nursing home. Findings from the rural municipalities indicate that those communicating with the hospital on transfer of patients were not always familiar with the patients but had the opportunity to refer them on to familiarised health personnel to better secure information continuity. Informants from rural municipalities said they knew many of their users, and it appears that a smaller environment resulted in better familiarity among staff, patients and next of kin. One participant, a ward leader, explained the process:

> I can receive a telephone call about a patient ready for discharge, or a 24-hour warning. It can arrive at reception or to the homecare nurse. If reception receives the call during the day they then place the call to the individual they think is the best receiver.
>
> **Ward leader, nursing home, rural municipality**

According to the informants, there is a great need for reliable communication between hospital and municipalities in connection with discharge of patients. They stated that communication occurs in different ways. Telephone, fax and EME were used. EME is a system enabling hospital and municipality to exchange patient information electronically. EME was in place between the university hospital and the urban municipality. The rural municipalities and the county hospital had not yet taken EME into use and instead used faxes for written dissemination of patient information. Information could be sent by telephone and fax, but when EME was not in place, organisations were dependent on documentation from the hospital being sent directly along with patients, to ensure that late changes were included. Informants from both rural municipalities described this process. Informants from the county hospital and associated municipalities largely believed that coordination and information flow worked well, and they observed that, as an extra security, they delivered information with the patient, sometimes with the help of next of kin, in addition to fax and telephone.

In the large municipality, lines of communication were somewhat simpler, because all notices of discharge-ready patients went to a Patient Service Unit. When a patient was discharged to a ‘transition ward’, there was additional contact between the department and the hospital. The results reveal that there are differences in how the flow of information proceeds between the hospital and the municipality, and these differences depend on the organisation and on whether EME has been taken into use. In the three municipalities, Coordinating Units functioned differently, and EME was only available in the urban municipality and at the urban university hospital. This has consequences for how healthcare personnel communicate with each other.

Figure 10.1 presents information flow between the two hospitals and municipalities, including obligatory notifications in accordance with the sub-agreement regarding hospital discharge of patients. There is a difference between municipalities with and without a Patient Service Unit.

When Is a Patient Ready for Discharge from Hospital?

It is the hospital’s obligation to determine when a patient is ready for discharge to the municipal health service, according to the agreement on discharge of somatic patients. The doctor at the ward is responsible, but result shows that the decision is taken in collaboration with the nurses. A participant from the county hospital reported engaging in discussion with the doctor to arrive at agreement about when the patient is ready for discharge. The doctor considered the patient’s medical status, and the nurse evaluated the patient’s level of functioning and need for training with reference to the municipality. Nurses decided whether notification of completed treatment was provided in accordance with the agreement.

Should a hospital fail to comply, the municipality could protest, and there have been cases where the hospital has had to retain the patient an extra day because of lacking notice or for failure to conform to notice requirements. In individual cases, disagreement can arise between the municipality and the hospital about whether the patient’s treatment is complete, and this can lead to conflict regarding whether the municipality is obliged to pay for the patient while still being hospitalised in accord with the agreement. Informants observed that it could be difficult to decide when the

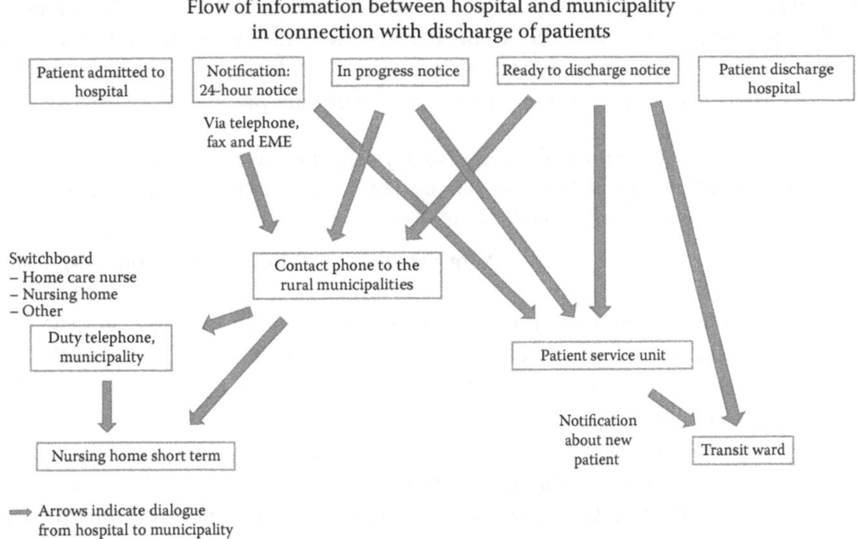

FIGURE 10.1 Flow of information between hospital and municipality in connection with discharge of patients.

patient's treatment is complete, because the health condition of the patient changes. Complications can arise which change the date of discharge, and the hospital needs to remain in regular dialogue with the municipality to keep the discharge date updated.

Informants noted that next of kin are important. Because of illness, medication, cognitive impairment and impairments of vision and hearing, older people often wish next of kin to be thoroughly informed of the condition of their health, because they are important assistants in the follow-up of services and of further medical help. Informants pointed to patients with dementia or cognitive impairment as the most vulnerable to rapid hospital admittance and discharge and to transfer between departments and organisations. According to the informants, discharge from hospital to short-term wards can occur quickly.

Our findings indicate that employees in administrative and managerial positions viewed the patient coordinator of the Patient Service Unit as important for both patient and next of kin and for the hospital and nursing home. The informants observed that quick transfers could make the patient uncertain and insecure about the situation and about further services and care. The patient coordinator has an important and challenging role in cooperating with the hospital to ensure coordination of services so that patients experience predictability. A number of informants cited examples of working across departments in order to find good solutions for the patients.

> There have been some especially vulnerable patients that have been moved twice. First been moved to the transition ward and then moved further to the intended place … but we have spoken about this, we have an oral agreement in a sense … that especially vulnerable patients shall be moved the fewest times possible.
>
> **Coordinating section, urban hospital**

The capacity of the municipality to receive patients on completion of treatment has significantly improved since the Coordination Reform was instituted, according to all informants. Informants from both hospitals wondered how this was possible. Their experience was of sending patients off earlier than previously.

> Previously they stayed perhaps ten to twelve days after they were finished and received their rehabilitation time in here, and so you are healthier when you leave. But now notice is given immediately when they are ready.
>
> **Administrative nursing leader, county hospital**

Informants from all municipalities confirmed the attitude that patients ready to discharge should be received. In cases of very sick patients where it is not ethically or professionally defensible to move the patient to a temporary location (transition ward, urban municipality), but where the patient must be transferred directly to the correct placement, the municipality may pay for the patient to remain in hospital. In some cases, the hospital also defers registering the patient as fully treated to spare the patient an extra transition. This occurs in collaboration with the municipality when the municipality can make a suitable offer within a short time. As a rule, responsibility for payment follows governing agreements, but collaboration to safeguard the rights of especially vulnerable patients occurs regularly, according to informants.

To avoid penalty fees, the urban municipality organised a transit ward to receive patients when ready to discharge. The rural municipalities sometimes took two patients in one room rather than one, and they used rooms not planned for patients to avoid paying fees.

There is a major convergence of views from the informants in hospitals and municipalities that economy is a concern for both organisations; financial concerns influence the hospital's decision to indicate discharge readiness, and for the municipality's decision to notify discharge readiness before the payment requirement would come into effect.

Competence and Experience

Most of the informants were nurses working in different management or administrative positions in hospitals or municipalities. They reported that nurses in different positions were occupied within coordination work internally and across organisations.

Experienced healthcare personnel are important to achieve continuity of care and service with regard to discharge of patients, according to informants. One of the informants from the county hospital, who had the responsibility for discharge of patients, noted that adequate time, experience, competence and stability among nurses are important factors contributing to secure discharges that safeguard continuity of services. Several nurses at the county hospital had been at the department for more than 20 years. As a result, their superiors felt secure that the nurses knew the requirements and followed them while also maintaining clinical professionalism.

The informants from a small rural municipality that was often in communication with the hospital about discharge of patients explained how important it was to be able to ask the right questions to hospital staff, to form a correct picture of patients and to clarify health conditions and level of functioning. She related the following:

> We have these forms we use, so when I pick up the telephone I am very concentrated. So I think that this can be irritating for them when they get me in conversation … they know then it is thorough. There must not be much room for doubt afterwards.
>
> **Ward leader, rural municipal nursing home**

The informant from the transition department in the urban municipality, which received all its patients from the hospital, was very concerned with the competence of those in dialogue with the hospital. The facility used specially trained nurses in these roles. The ward recognised the importance of clinical skills in properly caring for patients coming from hospital. The participant's experience was that good communication and clinical skills are necessary to ensure continuity of services for patients. Informants mentioned that cooperation between health personnel in different organisations is time-consuming and complex work. At both hospitals, informants stated that lack of time is most often the cause of deficiencies at discharge. They also observed that direct work with patients should be prioritised ahead of planning and discharge work. In busy periods, work is done using overtime or performed less thoroughly.

Informants in different hospital wards and municipalities told of a fast work tempo and of expectations for a good patient flow through the system. For the municipalities, this meant receiving patients ready to discharge immediately after they finished with hospital treatment. Informants from all municipalities observed that the need to provide for the patient's best interests, while at the same time meeting economic demands and ensuring patient flow, could be a conflictual experience. On the one hand, the employer desires efficiency in the use of resources, but the wishes of the patient and the next of kin may contradict this aim.

DISCUSSION

This study reports that the new agreements between the hospital and municipality health services have improved coordination between the organisations. There are concerns about shorter hospital stays and quicker transfers of patients from hospital to municipalities due to the discharge agreement and, in particular, about the consequences for elderly patients. There are differences between rural and urban municipalities and hospitals in how they have organised service coordination, their systems for information exchange and in the extent to which they have implemented EME. Experienced health personnel are important to ensure coordination and continuity of care.

Development of Agreements: Management Continuity

Management continuity involves both political health guidelines and the organisation of daily care for the patient (Hellesø 2012). Informants' satisfaction with discharge agreement is evidenced in the interviews. According to the informants, a more explicit delegation of tasks now occurs alongside established routines, leading to increased and earlier dialogue between the organisations. They believed the agreements provide predictability and guiding limits. A clearer delegation of work tasks between the administrative levels following implementation of the Coordination reform was also reported by Martens and Veenstra (2015).

It is demanding to negotiate new agreements and implement them in large organisations like hospitals and municipalities. Our results show that a thorough job has been done in this respect by disseminating information on the agreements and how to put them into practice in the organisations. Although work with agreements has been initiated and managed at the system level, work has been done to involve managers and personnel in different kinds of work groups. Findings indicate that, with the implementation of the new agreements, a feeling of greater equity emerged between the hospital and the municipality. Both parties experienced an evening out of the balance of power because of work with the agreements. Respect for co-workers, good structures of coordination and a balance of power are important factors that affect coordination (Øgar and Hovland 2004).

Coordination of services between health personnel in different organisations and units of care is a comprehensive and demanding task for nurses (Hellesø et al. 2005). Research has shown that when patients are moved, it is largely the nurses who are in contact with each other across and within organisations (Kirsebom et al. 2012; Moore 2012; Olsen et al. 2013). The way in which nurses solve problems involved in transfers from hospital to municipality healthcare service can make a difference in patients' experience of continuity (Hellesø 2012).

According to the informants in this study, nurses in hospitals and municipalities have an important coordinating task. Orvik (2006) referred to nurses as the glue of the organisations. Nurses contribute to holding a complex system together, a task other health professions take for granted. Nurses' many organisational tasks can compromise their time for patient contact, undermining the opportunity for the development of professional competence in their own clinical areas (Orvik 2006). The nurses interviewed here noted that coordination across organisations was time-consuming and that this had increased with the new agreements and consequences thereof, especially the obligatory notification and more frequent dialogue between the hospital and the municipality. The data do not give grounds for claiming that there are differences between rural and urban hospitals regarding increased workload.

Haggerty (2003) stated that management continuity is very important for patients with chronic or complex diseases that require services from several providers as well as for older persons discharged to short-term stays in nursing homes. According to Haggerty (2003), continuity is achieved when services are delivered in a complementary and timely manner. Informants in all municipalities reported that they were concerned about penalty fees for discharge-ready patients because rapid discharges can threaten the continuity of service due to lack of time for discharge planning. The economic incentive seems to work in accord with the purposes of the Coordination Reform. Informants noted that they had almost no patients ready to discharge in hospital. Starting in January 2012, change came about quickly. Municipalities have followed the requirements and taken on patients. This is in line with the intention emphasising the economic responsibility of the individual enterprise (Report to Parliament [No. 47] 2008–2009). However, shorter hospital stays and rapid hospital discharges can be a particular challenge for the older patients (Martens and Veenstra 2015).

Information Flow Has Significance for Coordination: Informational Continuity

The findings show a variation in how information flows between the hospital and the municipality depending on the organisation and on whether EME has been taken into use. The rural municipalities still had neither EME in hospitals nor a well-functioning Coordinating Unit, informants said. By contrast, the urban hospital and municipality had both in place. Informants from both the county hospital and the rural municipalities were optimistic about reduced workload in relation to exchange of information and more secure flow of information after the introduction of EME. Simple access to and exchange of relevant information is decisive for achieving good coordination and to ensure secure, proper treatment of patients (Hellesø 2012; Jeffs et al. 2013).

Continuity of information means that the healthcare personnel responsible for the patient are adequately informed about treatment, prognosis and further plans for treatment/care in both organisations, and the patient must also be so informed (Haggerty et al. 2003). Written information on the transfer of patients to short-term placements is a requirement in accordance with sub-agreement on discharge of patients, whether by fax or EME (Report to Parliament [No. 47] 2008–2009). When older patients are discharged from the hospital, documentation tends to focus on medical conditions, but it is also important to share knowledge of the patient's values and preferences (Haggerty et al. 2003). Our data do not support the claim that patients in small municipalities lack documentation to a greater extent than those in larger ones when discharged to short-term nursing home care, even though these small municipalities have no EME.

Informants related that patient and next of kin often bring along information from the hospital and are thereby messengers between the different organisations and responsible for safeguarding the continuity of information (Hellesø 2012; Hellesø et al. 2005). That the information exchange is performed in a number of different ways creates challenges, in particular for the hospitals that must deal with different procedures in different municipalities, as well as for both types of organisation where the use of fax and telephone is found to be cumbersome and laborious. Healthcare personnel in hospitals and municipalities maintained that they used a lot of time on information exchange (written and conversations) in order to ensure continuity of information.

The Vulnerable Patient: Relational Continuity

Relational continuity refers to the personal relationship between an individual patient and professionals, built on shared experience and interpersonal trust (Guthrie 2008; Haggerty et al. 2003). Elderly patients with chronic diseases traverse organisations, going from specialist to primary healthcare, which involves many different professionals. As primary healthcare has grown, the primary healthcare team has expanded, especially with nurses, who have also extended their roles (Guthrie 2008).

The informants involved with the Patient Service Unit reported that this system secures cooperation and continuity of care. The patient coordinator can become familiar with the patient and next of kin as they follow them through the different phases of illness and changing needs. Care for the individual and care delivered over time by the same healthcare professionals both ensure relational continuity (Haggerty et al. 2003). A good relationship to the patient coordinator, according to interview informants, facilitates provision of appropriate services and contributes to the patient's greater confidence in the system.

Relational continuity can lay the ground for how patients experience the quality of services. In this study, we do not have the voice of the patients and do not know if the patients experience relational continuity. Haggerty (2003) states that the experience of continuity may differ for the patients and the providers, which is a challenge in the evaluation of continuity.

In the smaller rural municipalities, services are not organised with a patient service unit and patient coordinators. In order to receive fax and telephones from the hospital, they use their existing services in nursing home, home care and switch board. In the smaller rural municipalities, people are more acquainted to each other, and often there are just one or two nursing homes, with which both the patients and the next of kin feel familiar. When patients and next of kin know or are familiar with healthcare personnel, they can have a sense of predictability and security, despite having no permanent healthcare personnel/patient coordinator to which to relate. In contexts where there is little expectation of establishing relationships with multiple healthcare personnel, as in nursing home care services, a stable core of personnel can give the patient a sense of stability and predictability (Haggerty et al. 2003). Informants in the rural municipalities indicated that they knew many of the patients personally, and this may lead to a sense of stability and predictability for the patients.

LIMITATIONS OF THE STUDY

Informants in this study were administrative personnel and managers. Managers have been involved in producing, disseminating and implementing the agreements at the departmental level. Administrative personnel and managers can be loyal to management and feel that they represent the service system; they may therefore find it difficult to express negative views in an interview (Glouberman and Mintzberg 2001). Had nurses been interviewed who experienced daily patient contact, different views on the coordination and transfer of elderly patients to the municipality healthcare service may have emerged. Further research focusing on the perspective of patients and next of kin will contribute valuable viewpoints on their experiences with continuity and coordination of care.

CONCLUSION

This study has investigated healthcare personnel's perceptions of coordination of healthcare services between the hospital and the municipality, especially when older patients are discharged to municipal short-term placements, in two counties in Norway. Data analysis revealed four themes of importance for the coordination of health services: *implementing new agreements, organisation of services and information flow, when is a patient ready for discharge from hospital?* and *competence and experience in health personnel.* The work with implementation of new agreements has involved managers, administrative personnel and clinical personnel at different levels of the organisations, as well as a range of different work methods. The agreement has led to a more explicit delegation of tasks and earlier dialogue between the organisations. There is variation among organisations regarding how they organise for coordination and how information flows between the hospital and the municipality. Nurses in hospitals and municipalities have an important, but time-consuming, role in coordinating services across organisations.

REFERENCES

Aase K, Laugaland AK, Dyrstad ND, Storm M. Quality and safety in transitional care of the elderly: The study protocol of a case study research design (phase 1). 3(8), 1–9. *BMJ Open* 2013; e003506. doi: 10 1136/bmjopen-2013-003506.

Bergmo TS, Ersdal G, Rødseth E, Berntsen G. Electronic messaging to improve information exchange in primary care. *The Fifth International Conference on e Health, Telemedicine, and Social Medicine*, Nice, France, 2013.

Dahl U, Steinsbekk A, Jenssen S, Johnsen, R. Hospital discharges of elderly patients to primary health care, with and without an intermediate care hospital – A qualitative study of health professionals' experiences. *International Journal of Integrated Care*, 2014; 14, 30.

Danielsen B og Fjær S. Erfaringer med å overføre syke eldre pasienter fra sykehus til kommune. *Sykepleien forskning*, 2010; 5(1), 28–35.

Det kongelige Helse- og Omsorgsdepartementet (The Norwegian Ministry of Health and Care Services). (2011–2015). *Nasjonal helse og omsorgsplan* (Meld. St.16). Oslo, Norway: Departementet.

Glouberman S, Mintzberg H. Managing the care of health and the cure of disease – Part 1: Differentiation. *Health Care Management Review* 2001; 26(1), 56–69.

Graneheim UH, Lundman B. Qualitative content analysis in nursing research: Concepts, procedures and measures to achieve trustworthiness. *Nurse Education Today* 2004; 24, 105–112.

Guthrie B. Why care about continuity of care? *Nzfp* 2008; 35(1), 13–15.

Haggerty JL, Reid RJ, Freeman GK, Starfield BH, Adair CE, Mckendry R. Continuity of care: A multidisciplinary review. Education and debate. *BMJ* 2003; 327 (7425), 1219–1221.

Hellesø R. Fra helsestyrt til pasientdrive informasjons- og koordineringsansvar? I: Melberg, H.O., Kjekshus, L.E (red), *Fremtidens helse-Norge*. Bergen, Norway: Fagbokforlaget Vigmostad og Bjørke AS, 2012; pp. 149–166.

Hellesø R, Sørensen B, Lorensen M. Nurses' information management across complex health care organizations. *International Journal of Medical Informatics* 2005; 74, 960–972.

Helsedirektoratet. Coordinating unit for habilitation and rehabilitation. http://www.helsedirektoratet.no/helse- og- omsorgstjenester/habilitering-rehabilitering/_(accessed 8 November 2012).

Jeffs L, Lyons R, Merkley J, Bell C. Clinicians' views on improving inter-organizational care transitions. *BMC Health Service Research* 2013; 13, 289. http://www.biomedcentral.com/1472-69631.

Kirsebom M, Wadensten B, Hedstrøm M. Communication and coordination during transition of older persons between nursing homes and hospital still in need of improvement. *Journal of Advanced Nursing* 2012; 69(4), 886–895.

Kvale S, Brinkmann S. *Det kvalitative forskningsintervju*, 2nd edn. Oslo, Norway: Gyldendal akademisk, 2009.

Laugaland AK. *Transitional Care of the Elderly from a Resilience Perspective.* Doctoral thesis, Faculty of Social Sciences, University of Stavanger, Stavanger, Norway, 2015.

Laugaland AK, Aase K, Barach P. Addressing risk factors for transitional care of the elderly – Literature review. In Albolini, S. et al. (eds.), *Healthcare Systems Ergonomics and Patient Safety.* London, U.K.: Taylor & Francis Group, 2011.

Laugaland AK, Aase K, Barach P. Interventions to improve patient safety in transitional care – A review of the evidence. *A Journal of Prevention, Assessment and Rehabilitation,* 2012; 41(Suppl. 1), 2915–2924.

Lov om kommunale helse og omsorgstjenester (Law of primary health and care services) m.m., 24. juni Nr. 30. 2011. I:Syse, A (redaktør) *Lovsamling fra helse - og sosialsektoren 2012–2013.* Oslo, Norway: Gyldendal juridisk; edition 22, 2012, pp. 958–988.

Lov om spesialisthelsetjenesten (Law of specialist care) m.m., 2. juli. Nr.61.1999. (Endret 18. desember 2009) I:Syse,A (redaktør) *Lovsamling for helse - og sosialsektoren 2012–2013.* Oslo, Norway: Gyldendal juridisk; edition 22, 2012, pp. 576–590.

Martens CT, Veenstra M. Samarbeidsavtaler mellom helseforetak og kommune. Fra dialog til avviksmelding? NOVA Velferdsforskningsinstituttet Høgskolen i Oslo og Akershus, rapport no. 9, 2015.

Moore MS. The European handover project: The role of nursing. *BMJ Quality and Safety,* 2012; 21(suppl. 1), 6–8.

Øgar P and Hovland T. *Mellom kaos og kontroll. Ledelse og kvalitetsutvikling i kommunehelsetjenesten,* 1st edn. Oslo, Norway: Gyldendal akademiske, 2004.

Olsen RM, Hellzen O, Enmarker I. Nurses' information exchange during older patient transfer: Prevalence and associations with patient and transfer characteristics. *International Journal of Integrated Care* 2013, 13(1), 1–9.

Orvik A. *Organisatorisk kompetanse – i sykepleie og helsefaglig samarbeid.* Oslo, Norway: J.W. Cappelen forlag A.S. 2006. 3. Circulation, 2004.

Report to Parliament No. 47 (2008–2009). Coordination reform: Proper treatment – At the right place and at the right time. Available at: http://www.regjeringen.no/upload/HOD/Dokumenter%20INFO/Samhandling%20engelsk_PDFS.pd.

Romøren IT, Torjesen OD, Landmark B. Promoting coordination in Norwegian health care. *International Journal of Integrated Care* 2011; 11, e127.

Samhandlingsavtaler. Coordination agreements between health enterprises and municipalities. Available at: http://www.helsestavanger.no/no/FagOgSamarbeid/Samhandling/Documents/Avtaler/Stavanger/J%20-%20Stavanger%20LSA%20-%20Delavtale%20nr.%205; PDF – https://statistikk.samhandlingsbarometeret.no/webview/ (accessed 25 August 2015).

Storm M, Siemsen IM, Laugaland K, Dyrstad DN, Aase K. Quality in transitional care of the elderly: Key challenges and relevant improvement measures. *International Journal of Integrated Care* 2014; 14(2), 1–15.

Thagaard T. *Systematikk og innlevelse – en innføring i kvalitativ metode,* 3rd edn. Bergen, Norway: Fagbokforlaget, 2010.

Tjora A, Tøndel G. *Fremtidens pasient* (*The future patient*), in Melberg H, kjekshus LE (red), *Fremtidens helse- Norge.* Bergen, Norway: Fakbokforlaget Vigmostad og Bjørke AS, 2012, pp. 105–124.

11 Leading Quality and Patient Safety Improvement in Norwegian Hospitals

Inger Johanne Bergerød and Siri Wiig

CONTENTS

INTRODUCTION

The Norwegian healthcare system is undergoing significant reform including an increasing focus on 'quality', of which safety is an important component (IOM, 2001; Wiig et al., 2014). The government's priority is to build a high-quality and patient-centred healthcare service that has a better distribution of power and influence between the system and the patient. In addition to strengthening the user role, the government has three focal areas to enhance quality and safety: (1) creating and establishing better systems, (2) strengthening leadership engagement and (3) building a reporting and learning culture to reduce adverse events (Ministry of Health and Care Services, 2008–2009, 2011–2012, 2012–2013; The Norwegian Directorate of Health, 2005).

As part of this increased attention to quality and safety, Norway launched its first national patient safety campaign in 2011, followed by a permanent patient safety programme in 2014. Leadership is one of the top priorities of the programme. Leadership and hospital boards play an important role in the coordination of complex hospital organisations and are often considered key links in these systems (Donaldson et al., 2000; Goeschel et al., 2010; Künzle et al., 2010). Research shows that implementation and maintenance of new methods are a demanding undertaking that requires comprehensive leadership, engagement, prioritising and enthusiasm (Bate et al., 2008; Frankel et al., 2008). Despite increased research into the role of leadership, there is still limited knowledge on how and in what ways leadership affects quality and safety in healthcare systems (Hockey and Bates, 2010; Levey et al., 2007).

The aim of this chapter is to explore the role of leadership in organisational and cultural factors related to quality and safety in hospitals. We compare the organising and leadership processes in two Norwegian hospitals. The following research question guided this study: How is leadership influencing quality and patient safety work in two Norwegian hospitals?

The study is part of the Norwegian case study in the European FP7-funded project 'Quality and Safety in European Union Hospitals: A research-based guide for technical implementing best practice and a framework for assessing performance', QUASER, 2010–2013 (Robert et al., 2011).

THEORETICAL APPROACH

To understand how hospitals are organising for quality and safety, and the role of leadership in this process, we applied Organising for Quality (Bate et al., 2008) as our theoretical framework. The evidence-based framework uses case studies in seven prominent hospitals in the United States and Europe to identify six challenges to achieving and sustaining quality of care. The original framework illustrates an integrated leadership perspective. In this chapter, we added two more challenges – the contextual challenge and the senior leadership challenge that were also identified in the QUASER study (Fulop et al., 2012).

The eight challenges applied in the theoretical framework are as follows:

1. *Structural* – organising, planning and coordinating quality efforts
2. *Political* – addressing and dealing with the politics of change surrounding any quality improvement effort
3. *Cultural* – giving 'quality' a shared, collective meaning, value and significance within the organisation
4. *Educational* – creating a learning process that supports improvement
5. *Emotional* – engaging and mobilising people by linking quality improvement efforts to inner sentiments and deeper commitments and beliefs
6. *Physical and technological* – designing physical systems and technological infrastructure that support and sustain quality efforts
7. *Contextual* – responding to social, political and contextual factors outside and inside the organisation
8. *Senior leadership* – evaluating all leadership challenges in the framework related to quality and patient safety (Figure 11.1)

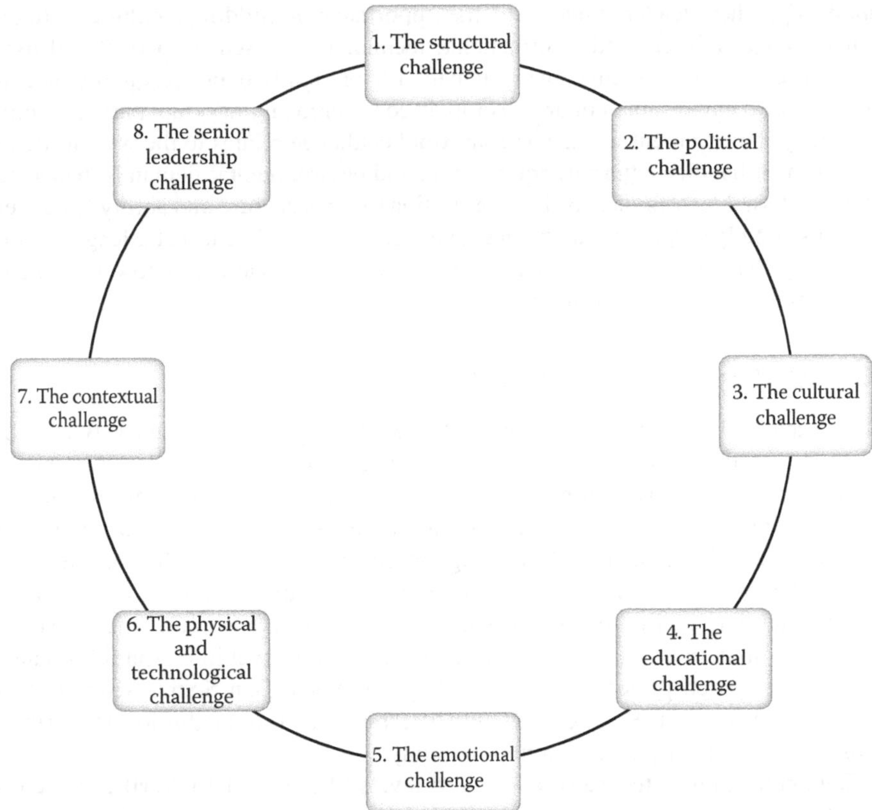

FIGURE 11.1 Eight leadership challenges. (Based on Bate, P. et al., *Organizing for Quality: The Improvement Journeys of Leading Hospitals in Europe and the United States*, Radcliffe, Oxford, U.K., 2008.)

The integrative leadership perspective in our theoretical framework implies a variety of approaches and priorities to meet the challenges. Within the structural challenge, we find *strategic leadership* involving a strong vision, strategic goals and discipline around the focus on quality and safety. Within the political challenge, we find *politically credible leadership* in which leaders with authority, negotiating skills and ability respond to quality and patient safety challenges with seriousness and conviction. Clinical empowerment is important in this challenge as is the ability to align messages to the medical specialties in the hospital. In the cultural challenge, we find value and *symbol-based leadership* in which leaders cultivate a culture that is dedicated above all to quality and patient safety. In the educational challenge, we find *educational leadership* of people who work with best practice, learning and development in quality and patient safety. The emotional challenge focuses on *inspirational leadership* where enthusiastic leaders see quality and patient safety as a 'goal' or a project. They suggest ideas, methods and motivations for the implementation and maintenance of quality and safety improvement throughout the organisation. The physical design and technology challenge is related to *technological-based leadership*, where leaders understand the importance of building a culture of functional, ethical, clinical and informational technologies based on patient and user experiences. In this challenge, the systematic use of quality data (e.g. quality indicators), physical environment or architecture is considered an important part of quality and safety improvement work. The contextual challenge relates to the way in which management handles external requirements and environmental demands, balanced against internal factors such as the organisation's size, structure and ability to deliver in terms of budget, quality, safety and patient satisfaction. The last challenge, senior leadership challenge, is an overall assessment of leadership attention towards organising for quality and patient safety.

METHODOLOGICAL APPROACH

This chapter is based on the Norwegian QUASER study of two hospitals (Bergerød, 2012; Bergerød and Wiig, 2015; Wiig et al., 2013, 2014). The Norwegian hospitals were selected on the basis of performance on five national quality indicators (average over the last 3 years). Hospital A was designated as 'high performing' and Hospital B was designated as 'developing' based on the selected indicators (Burnett et al., 2013; Wiig et al., 2011). Both hospitals were studied at the meso (organisational) and micro (ward) levels. The study included two areas at the micro level in Hospital A (maternity, cancer) and one area at the micro level in Hospital B (maternity). The research project was approved by the Norwegian Social Science Data Services (5 May 2011, Ref. 26636), and all respondents provided informed written consent prior to their participation in the study.

Data collection at the meso and micro levels (April 2011 to April 2012) consisted of semi-structured interviews (97), focus group interviews (2), interview with patient representatives and patient Ombudsman (Hospital B), shadowing of healthcare professionals and observation (45 h) and documentary analysis (strategies, mission documents, board reports, user committee reports, method books). A common interview guide was developed based on the theoretical framework (Figure 11.1).

All interviews were transcribed and uploaded in Nvivo and analysed according to a predefined codebook according to the eight challenges in our theoretical framework (structure, politics, culture, emotions, education, physical and technical, external demands and leadership) (Robert et al., 2011; Wiig et al., 2014). In this chapter, we focus on the leadership aspect related to the eight challenges (Bergerød, 2012).

LEADING QUALITY AND SAFETY IMPROVEMENT IN HOSPITALS

In the following, we will present results from the Norwegian QUASER study according to the eight leadership challenges (Figure 11.1). Under each challenge, we provide examples in the form of quotes from Hospitals A and B. Table 11.1 gives an overview of main contextual factors for the two hospitals.

TABLE 11.1
Contextual Factors for the Two Case Hospitals

Context	Hospital A	Hospital B
Location	Small town in rural Norway	Large city in Norway
Population	110,000 inhabitants	260,000 inhabitants
Hospital	Central hospital	University hospital
	Small hospital	Large hospital
	Treats 140,000 patients annually	Treats 600,000 patients annually
Employees	2,200 employees	11,500 employees
Annual budget	1.4 billion NOK	8.4 billion NOK

STRUCTURAL CHALLENGE

Hospital A is a small hospital with a hierarchical organisational structure. Hospital B is a large teaching hospital with a flat organisational structure. Formal responsibility for quality and safety relies in the management line organisation. Both hospitals have established a mandatory quality committee and a patient safety committee. Hospital A's quality committee is highly respected, and all division directors and top management sit on it. The latter is a senior management decision made to pinpoint the role of leaders in working on quality and safety. As a consequence, Hospital A's quality committee is active in quality improvement. In contrast, Hospital B's quality committee is characterised as a 'sleeping' committee. It has rather little prestige, and none of the committee members are clinicians with authority.

Organisational support for line managers, with regard to the efforts to improve quality and safety, differs significantly. Hospital A has a small number of quality staff to support the line managers in improvement efforts and projects. All improvement projects follow the same methodological approach. The rationale for the small staff and the one improvement method is to make the line managers responsible for ensuring quality and safety in their own department. Responsibility should not be handed over to staff outside the unit, but it can be supported by professionals who are well versed in the common improvement method. Hospital B has a relatively large patient safety section with the expertise and educational resources to support the managers across the hospital. This organisational section is central in supporting the line managers in working on quality and safety improvement. It acts as an internal knowledge centre with dedicated educational resources. The result shows that the patient safety section is successful in helping line managers exercising their formal responsibility for quality and safety within their department (Table 11.2).

POLITICAL CHALLENGE

The senior management at both hospitals works politically but in different ways to enact change. Hospital A's senior managers combine 'top-down' and 'bottom-up' approaches to quality and safety improvement and change. In practice, our result

TABLE 11.2
The Structural Challenge

Hospital A	Hospital B
About a common improvement method for working on quality and safety improvement: *We aim for good results in all the improvement projects included in the improvement program. We also aim for developing a comprehensive methodology that has organization wide support. Our methodological approach takes employees seriously and we use them as strategic consultants towards the goal of securing equal rights to high quality health care services* (Document: Application to be a pilot hospital for improvement programme).	About the importance of formal quality and safety committee: *This committee [the quality committee] has been 'sleeping'. It has been argued that the top management group should be included in committee* (Administrative staff, Patient safety section). *The patient safety committee is constituted by highly qualified clinicians holding a high credibility across professional disciplines. The participants are very competent and represent different professional disciplines. They meet on a monthly basis and receive cases from the patient safety department* (Administrative staff, Patient safety section).

shows that 'top-down' leadership seems to work well in small organisations. This is evident in Hospital A because senior managers are aware of their own role in quality and safety improvement and of long-term organisational development that is attentive to building a shared organisational cultural within the hospital. These senior managers do not think that they need to spend time persuading employees to participate in improvement activities or to support changes imposed from the external environment. The senior management trusts the organisational culture to assist with the adoption and implementation of changes.

In Hospital B's flat organisational model, clinical empowerment is a key in the bottom-up leadership approach to quality and safety improvement processes. Senior managers argue that the clinical empowerment is necessary to enlist employees in improvement efforts. Moreover, the senior managers need the managerial skills to deal with power issues and negotiations of change. The latter implies that the senior managers spend much time 'selling the message' to get key clinicians on board as part of the change process.

The political challenge of change is present in both hospitals. For example, the maternity wards underwent structural changes to cut costs. Hospital A merged the maternity and gynaecology departments. Hospital B downsized its staff at the expense of obstetrical care. The changes created resistance and power conflicts in both hospitals, because healthcare staff was given new tasks in Hospital A (midwives and nurses were assigned tasks across maternity and gynaecology) and because the staff resources and expertise were transferred between wards and allocated to patients with the highest risk potential in Hospital B (from training in breastfeeding in the post-natal ward to emergency situations on the maternity ward). Despite resistance, both hospitals accepted the imposed changes (Table 11.3).

TABLE 11.3
The Political Challenge

Hospital A	Hospital B
About the consequences of merging the maternity and gynaecology departments:	About amending processes and economics:
I assume that responsibility is imposed on us and we have a feeling of not being sufficiently competent. We are constantly striving. I feel it is assumed that we as nurses can work anywhere, but this is not the case. The demands on us are large in terms of both quality standards and knowledge needs. Should a doctor go from the general department to the surgical department and make his rounds there? This striving is very challenging (Nurse).	*Currently we talk about economy, yes, we still talk about that, but not economic aspects only. We talk a lot about professional development, patient quality, patient safety, how to improve patient pathways in an efficient manner and how to solve the targets specified in the letter of assignment, such as waiting lists, priorities such as deadlines related to discharge summaries* (Senior manager).

CULTURAL CHALLENGE

The two hospitals have different cultural traits. Hospital A is characterised by a coherent, open and non-punitive culture. Healthcare professionals show a collective commitment to the provision of high-quality and safe healthcare services. The organisation is relatively small and characterised by oversight and short distance between the shop floor and the senior managers. Senior managers are committed to building a common culture through language adjusted to the language used by the staff at the micro level.

Hospital B has a culture with strong professional groups and a strong professional identity. This implies that it can be difficult for the staff that is not leading clinicians to talk about incidents or to speak up, either formally or informally. Underreporting is perceived to occur frequently. A central cultural feature in the maternity ward at Hospital B is that healthcare staff (doctors/midwives) 'defines' quality for the patients (mother and baby), to give birth vaginally rather than through caesarean section. Patient experience is usually not on the agenda, and the culture is characterised by an informal professional competence hierarchy (Table 11.4).

EDUCATIONAL CHALLENGE

The hospitals differ in the ways in which senior managers work in relation to learning and education of healthcare professionals. Hospital A has had great challenges to meet the demands from the inner and outer context on how to educate and train its employees. Due to a difficult staffing situation with few specialists, there are limited opportunities to send employees to courses, conferences and professional development. Hospital A is very systematic in its handling of adverse events or incidents, despite a limited number of staff that works directly with such cases. All adverse events are reviewed by the quality staff and discussed in the quality committee in order to promote system-wide learning from adverse events.

TABLE 11.4
The Cultural Challenge

Hospital A	Hospital B
About creating a common culture through language: *I am not so dandy that I have to use all these fancy terms and concepts. So, I think we have to create our own language and use our own words. The underlying meaning is the same* (Senior manager). *An action plan is easier to understand than a project directive. Those are the things we are talking about* (Senior advisor).	About informal hierarchy in the department: *If you are «in» you can come away with anything, and people don't argue. Then, there is always someone who can't say a thing without being corrected. I think we have to improve the culture, especially among the doctors … I think it is very difficult to be in the specialist training program at this hospital. It is a long way to work your way up in this hierarchy. I have said it a number of times … I am so glad I did not get my specialist training at this hospital* (Physician).

Hospital B is characterised by an academic tradition and research environment where the staff has a strong focus on learning both from within the organisation and from the international research community. Senior managers in Hospital B emphasise the recruitment advantages of strong research traditions and being in the forefront of research. The patient safety section is central to learning from adverse events.

The role of professional education of nurses/midwives is a hallmark of the learning processes in both hospitals. They contribute to professional development and updating of procedures and organise simulation sessions to offer training in technical skills as part of the ordinary work on the wards, when time permits (Table 11.5).

TABLE 11.5
The Educational Challenge

Hospital A	Hospital B
About difficulties in prioritising education and training activities: *The manager tells us that we can attend scientific conferences. But the problem is that we can't go because of the staffing situation. There are often too few of us here. Sometimes it is almost impossible to run the department when key personnel are gone* (Senior physician). *The quality improvement work will always compete with other tasks taking place in a hospital* (Senior advisor).	About professional development as part of the recruitment strategy: *We do have many that graduate for a PhD degree here … it is a part of our policy to educate our own employees. That is why we try to facilitate a work situation allowing for doing a PhD while being employed here. This is part of our recruitment strategy* (Senior manager).

TABLE 11.6
The Emotional Challenge

Hospital A	Hospital B
About senior management enthusiasm: *Look to Jønkøping [a city in Sweden]» it said, and I thought, what...? Why not look to this county? What was so special about Jønkøping? Nothing! They did not do anything fabulous, but they did what they did in a systematic way and then they created a culture* (Senior manager, Hospital A).	About competence as a driving force: *... this department is considered a very good department with high professional status that prioritizes professional expertise. Professional expertise trumps everything.* (Senior physician)

EMOTIONAL CHALLENGE

The two hospitals show different abilities to instil positive feelings, inspiration and commitment to quality and safety. Despite external pressure for cost efficiency, mergers and downsizing and even threats of shutdown, the employees and the managers show a mobilising response to the 'ghost of change'. The merger of the maternity and gynaecology departments in Hospital A was not only frustrating for employees but also elicited a consistent striving and a collective enthusiasm for the delivery of high quality and safety. They wanted to show the outside world that they delivered sound services with high quality. Professional pride, idealism and professional enthusiasm characterise Hospital B, where improvement is driven by 'bottom-up' leadership. Opportunities for financing quality improvement projects from regional health trust and participation in national patient safety campaign are key contextual factors that created enthusiasm and awareness of quality and safety work in both hospitals (Table 11.6).

PHYSICAL AND TECHNOLOGICAL CHALLENGE

Both hospitals pay close attention to collecting and using various types of data to improve the quality and safety of their services. Hospital A began its systematic quality and safety work in 2005 by collecting, analysing and presenting relevant quality data in order to prevent what the senior manager perceived as a 'culture of assumptions' in which professionals and managers assumed they knew the quality performance of their organisation, but with no facts or evidence to support these assumptions. The senior manager appointed a dedicated person with strong analytical skills and Hospital A started systematising data on quality indicators and established mandatory reporting on specified quality domains. Based on this effort, Hospital A is earning top scores on the established national quality indicators developed by the Directorate of Health.

Hospital B's managers and staff experience physical infrastructure in terms of buildings and localities as a challenge to the provision of high quality and safety within the maternity section. For example, the lack of space in the maternity section,

TABLE 11.7
The Physical and Technological Challenge

Hospital A	Hospital B
About the lack of tools and methods for user involvement and collecting patient experiences: *I have a very bad conscience when I am thinking about user involvement in quality improvement. We are really bad at it and we know it. I really want to do something about it in practice and not just on paper. The regional health trust has developed an IT tool for the user survey and the purpose is that we are going to implement it here* (Senior manager).	About physical location as obstacle to high quality: *The least satisfying thing about working here is the conditions we offer to the women on the post-natal ward. We do not provide them with a single room and they can't bring the father to stay with them. We don't have a family-room here – which we are actually obligated to have. It is very unsatisfactory* (Head of department).

poor facilities for mothers and infants and a lack of co-location between maternity and children's clinic are physical challenges that interfere with the provision of maternity care services according to current quality standards. Both staff and hospital managers agree that the need to transport the newborns, which required intensive care, from the maternity section to another building located in the children's clinic, poses a serious quality and safety challenge. The co-location of these services has been discussed for years. Both hospitals lack the competence and tools to collect and use patient experience to improve service quality and safety (Table 11.7).

CONTEXTUAL CHALLENGE

Governmental and political attention to quality and safety has increased in Norway over the last decade, in terms of national strategies, patient safety campaign and programme, regulation and targets affecting senior managers in both hospitals. The increased external demands and targets related to mangers' attention to quality and safety, in addition to the increasing requirements of reporting on national quality indicators, all contribute to legitimising managers' strategic effort to improve quality and safety. One of the most important external conditions for both hospitals is the link to the regional health authority (RHA). The RHAs are responsible for the delivery of the services within the health region. Every year, each of the four RHAs hands over a letter of assignment to the local health trust within the regions. The RHA contributes with providing courses, organising conferences and collaboration in relation to the national patient safety campaign. Hospital A takes advantage of this collaboration as it contributes to developing the organisations' professional competence within improvement work. Within the area of patient experience, the RHA supports the hospitals in developing tools to document patient experiences. Such measures are requested by both hospitals and illustrate the need for competence and support to take advantage of patient experiences as an important input for the improvement of quality and safety (Table 11.8).

TABLE 11.8
The Contextual Challenge

Hospital A	Hospital B
About long-term attention towards national quality indicators:	About external governing principles:
If I go back in time and read reports from 2006–2007 and look at time of discharge summary before we established the new way of reporting, the figures were in the range of 40%, 20%, 60%. Today's figures show 80% and some departments have 100%. I do not think we would have achieved this goal without the new way to using data and establish requirements of reporting (Senior manager).	*For all health care professionals, professional pride and performing high quality practice according to professional standards appear as keys. Earlier, these standards were set by health care professionals themselves. Currently, this has changed and the standards for sound professional practice are defined by bureaucrats and politicians* (Clinical director).

SENIOR LEADERSHIP CHALLENGES

Senior managers at both hospitals ensure quality and safety by putting the topic on their agenda. This sets a clear and strategic direction for the hospital staff and underpins the message that quality and safety are equally important to operational measures. This contributes to building legitimacy for the required time and effort in addition to holding managers at all levels accountable for providing of high-quality and safe services. In Hospital A, the senior managers emphasise symbolic (quality and safety culture) and technological (systematisation of data) leadership to improve quality and safety. Since 2005, the senior managers have worked to build a common culture of continuous monitoring of quality data. This constitutes an important factor of Hospital A's commitment to quality and safety. In Hospital B, senior managers emphasise political and educational leadership as key factors of success. Here, leaders must have the ability to negotiate, involve and collaborate with prominent clinicians. Leadership processes within Hospital B depend on providing the employees with sufficient time for professional development and research activities as a key element to improve quality and safety (Table 11.9).

LESSONS LEARNT ABOUT QUALITY AND SAFETY LEADERSHIP PROCESSES

Leadership along with other organisational and cultural factors strongly influenced how quality and safety work was carried out in hospital settings. This study of two Norwegian hospitals showed that managers and their organisations had different responses and actions to overcome challenges in improving quality and safety. Figure 11.2 summarises the hospitals' emphasis on the eight leadership challenges using a score from 1 (low) to 10 (high).

TABLE 11.9
The Senior Leadership Challenge

Hospital A	Hospital B
About leadership philosophy: *We have worked extensively with organizational development, delegation and understanding of roles. I also think this has been important in terms of developing our quality improvement work […]. I think this is important because previously we were under the impression that we could just sit on the top and makes decisions* (Senior advisor).	About leadership philosophy: *A key characteristic of our organization is the large degree of delegation of power. In comparison, other comparable hospitals have a centralized power structure within the organizations* (Senior manager).

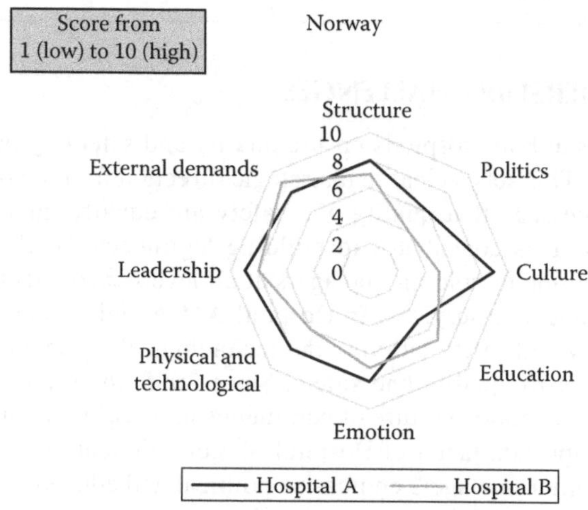

FIGURE 11.2 Overall assessment of Hospitals A and B on the eight leadership challenges.

In both hospitals, we find extensive senior manager commitment and a long-term perspective on quality and safety. Hospital A has integrated quality and safety improvement as an important part of daily operations without a special use of additional resources and staff in a quality department. Hospital A has become successful in creating a common organisational culture. Through a systematic use of quality data over time, the hospital has managed to achieve among the highest scores on national quality indicators.

Hospital B also pays considerable attention to the reporting and systematisation of data but has not scored as well as Hospital A on national quality indicators. This study makes no conclusion about the relationship between national quality indicators and quality results. However, the study shows that long-term commitment and involvement from the hospital senior management in quality and safety improvement are vital for how this is taken care of at the sharp end. Managers in both hospitals are sensitive to the outer context (indicators, letter of assignments). In addition, they are responsible for setting quality and safety on the agenda in the organisation and for the integration and adaptation of quality and safety improvements to the operation of the hospital. In other words, our results are similar to those of other studies, showing the importance of leaders in improving quality and safety (Frankel et al., 2008; Goeschel et al., 2010; Jha et al., 2010; Künzle et al., 2010; Leape et al., 2009; Levey et al., 2007). What is missing in the literature on quality and safety in healthcare services is the lack of comprehensive understanding of how different organisational, cultural and leadership factors interact in different clinical settings (Bate et al., 2008; Hockey and Bates, 2010; Krein et al., 2010). The latter is documented in this study through the two hospitals' quality and safety journey and response to the different leadership challenges, organisational and cultural factors. The study indicates that success on the leadership challenges cannot be attributed to a narrow focus on a single challenge such as structure or politics, but rather to a long-term approach to several challenges (Hospital A: structure, culture, leadership, emotions, physical/technical aspects; Hospital B: external requirements, leadership, structure, education).

Organisational size is an important explanatory factor for the successful quality and safety improvement work in Hospital A. The organisation is transparent with oversight; there is a short distance from the managers to the healthcare professionals at the shop floor and a good opportunity to build a common organisational culture. The cultural challenges differed between the two hospitals (Figure 11.2). The obvious leadership challenge in Hospital B is to improve a culture that is heavily shaped and influenced by leading clinicians and their definition of what is considered high quality and safety in healthcare services. Instruments other than 'bottom-up' initiatives focusing solely on clinical effectiveness should be implemented. International research literature (Coulter, 2011; Doyle et al., 2013; Robert and Cornwell, 2013) and governmental expectations (Ministry of Health and Care Services, 2011–2012, 2012–2013) point out the need for a culture that recognises the current patient role and considers patients and their experiences as one of several resources in quality and safety improvement efforts. This study concludes that involvement of patient or user experience needs to be given extensive attention by senior managers and other stakeholders in Norwegian hospitals.

Hospital B scores slightly better than Hospital A on the educational challenge (Figure 11.2). Although Hospital B has challenges associated with a learning culture because of strong professional roles, it has succeeded in facilitating professional development. This is of key importance for the quality and safety improvement processes at the hospital. Challenges such as external requirements related to the budget balance, cost efficiency and downsizing have not threatened the hospital's ability to prioritise professional competence development and updating. Through its prioritisation of continuous professional development, the hospital has also managed to establish a recruitment strategy that attracts talented professionals.

Few studies have explored the interactions among organisational, cultural and leadership factors at different levels in the healthcare system (macro–meso–micro) and their influence on quality and safety improvement practice in hospitals (Bate et al., 2008). The results are not generalisable as they are based on two hospitals with different sizes and functions. The results will still be relevant to other hospitals because the manners in which senior managers diagnose and approach different organisational and cultural factors are keys in ensuring quality and safety. Senior managers need to address these factors in a systematic process of 'self-diagnosis'. The focus and measures to improve practice will depend on organisational size, culture, competence composition, power relations and physical/technological conditions. The understanding of the organisational and cultural conditions and the role of leadership within each hospital is thus central, although expectations, requirements and regulations of the authorities are the same for all hospitals.

ACKNOWLEDGEMENTS

The study, 'Quality and Safety in European Union Hospitals: A Research-based Guide for Implementing Best Practice and a Framework for Assessing Performance (QUASER)', has received funding from the European Community's Seventh Framework Programme (FP7/2007–2013) under grant agreement n° 241724.

We wish to acknowledge the following members of the QUASER project:

Naomi Fulop, Glenn Robert, Janet Anderson, Susan Burnett, Charles Vincent, Susie Edwards, Heide Poestges, Kathryn Chales and Anna Renz, England; Roland Bal, Anne Marie Weggelar, Julia Quartz and Hester van de Bovenkamp, the Netherlands; Karina Aase and Christian von Plessen, Norway; Francisco Nunes, Sara Gomes and Alexandra Fernandes, Portugal and Boel Anderson Gäre, Johan Calltorp, Pär Höglund, Tony Andersson, Anette Karltun and Johan Sanne, Sweden.

We wish to thank all of the hospital employees and managers for taking their time to share their knowledge and experiences with us. Last, but not least, we thank the site managers at the hospitals who generously helped us organise the data collection.

REFERENCES

Bate, P., Mendel, P. and Robert, G. 2008. *Organizing for Quality: The Improvement Journeys of Leading Hospitals in Europe and the United States*. Oxford, U.K.: Radcliffe.

Bergerød, I.J. 2012. *Ledelse, kvalitet og pasientsikkerhet: sammenlignende case studie av to norske sykehus*. University of Stavanger, Stavanger, Norway.

Bergerød, I.J. and Wiig, S. 2015. Ledelse og pasientsikkerhet. In: Aase, K. (ed.), *Pasientsikkerhet – teori og praksis*, 2nd edn. Oslo, Norway: Universitetsforlaget.

Burnett, S., Renz, A., Wiig, S., Fernandes, A., Weggelaar, A.M., Calltorp, J., Anderson, J.E., Robert, G., Vincent, C. and Fulop, N. 2013. Prospects for comparing European hospitals in terms of quality and safety: Lessons from a comparative study in five countries. *Int J Qual Health Care*, 25, 1–7.

Coulter, A. 2011. Engaging patients in healthcare. Open University Press.

Donaldson, M.S., Kohn, L.T. and Corrigan, J. 2000. *To Err Is Human: Building a Safer Health System*. Washington, DC: National Academy Press.

Doyle, C., Lennox, L. and Bell, D. 2013. A systematic review of evidence on the links between patient experience and clinical safety and effectiveness. *BMJ Open*, 3, pii: e001570.

Frankel, A., Grillo, S.P., Pittman, M., Thomas, E.J., Horowitz, L., Page, M. and Sexton, B. 2008. Revealing and resolving patient safety defects: The impact of leadership WalkRounds on frontline caregiver assessments of patient safety. *Health Serv Res*, 43, 2050–2066.

Fulop, N., Walters, R., Perri and Spurgeon, P. 2012. Implementing changes to hospital services: Factors influencing the process and 'results' of reconfiguration. *Health Policy*, 104, 128–135.

Goeschel, C.A., Wachter, R.M. and Pronovost, P.J. 2010. Responsibility for quality improvement and patient safety: Hospital board and medical staff leadership challenges. *Chest*, 138, 171–178.

Hockey, P.M. and Bates, D.W. 2010. Physicians' identification of factors associated with quality in high- and low-performing hospitals. *Jt Comm J Qual Patient Saf*, 36, 217–223.

IOM. 2001. Crossing the quality chasm: A new health system for the 21st century. In: Committee on Quality Health Care in America (ed.), *Crossing the Quality Chasm: A New Health System for the 21st Century*. Washington, DC: The National Academies Press.

Jha, A.K., Prasopa-Plaizier, N., Larizgoitia, I., Bates, D.W. and Research Priority Setting Working Group of the World Alliance for Patient Safety. 2010. Patient safety research: An overview of the global evidence. *Qual Saf Health Care*, 19, 42–47.

Krein, S.L., Damschroder, L.J., Kowalski, C.P., Forman, J., Hofer, T.P. and Saint, S. 2010. The influence of organizational context on quality improvement and patient safety efforts in infection prevention: A multi-center qualitative study. *Soc Sci Med*, 71, 1692–1701.

Künzle, B., Kolbe, M. and Grote, G. 2010. Ensuring patient safety through effective leadership behaviour: A literature review. *Saf Sci*, 48, 1–17.

Leape, L., Berwick, D., Clancy, C., Conway, J., Gluck, P., Guest, J., Lawrence, D. et al. 2009. Transforming healthcare: A safety imperative. *Qual Saf Health Care*, 18, 424–428.

Levey, S., Vaughn, T., Koepke, M., Moore, D., Lehrman, W. and Sinha, S. 2007. Hospital leadership and quality improvement: Rhetoric versus reality. *J Patient Saf*, 3, 9–15.

Ministry of Health and Care Services. 2008–2009. Report No. 47 to the Storting (2008–2009) The Coordination Reform – Proper treatment – At the right place and right time.

Ministry of Health and Care Services. 2011–2012. Meld. St. 16 (2010–2011) Report to the Storting (white paper) Summary – National Health and Care Services Plan 2011–2015.

Ministry of Health and Care Services. 2012–2013. Meld. St. 10 (2012–2013) High Quality – Safe Services – Quality and Patient safety in the Health and Care Services.

Robert, G. and Cornwell, J. 2013. Rethinking policy approaches to measuring and improving patient experience. *J Health Serv Res Policy*, 18(2), 67–69.

Robert, G.B., Anderson, J.E., Burnett, S.J., Aase, K., Andersson-Gare, B., Bal, R., Calltorp, J. et al. 2011. A longitudinal, multi-level comparative study of quality and safety in European hospitals: The QUASER study protocol. *BMC Health Serv Res*, 11, 285.

The Norwegian Directorate of Health. 2005. Og bedre skal det bli – National Strategy for Quality Improvement in Health and Social Services 2005–2015.

Wiig, S., Aase, K., Von Plessen, C., Burnett, S., Nunes, F., Weggelaar, A.M., Anderson-Gare, B., Calltorp, J., Fulop, N. and For, Q.-T. 2014. Talking about quality: Exploring how 'quality' is conceptualized in European hospitals and healthcare systems. *BMC Health Serv Res*, 14, 478.

Wiig, S., Harthug, S., Von Plessen, C. and Burnett, S. 2011. Measuring quality and safety in Norwegian health care – Time for a change? In: Albolino, S., Bagnare, S., Bellani, T., Llaneza, J., Rosal, G. and Tartaglia, R. (eds.), *Healthcare Systems Ergonomics and Patient Safety 2011 – An Alliance between Professionals and Citizens for Patient Safety and Quality of Life*. London, U.K.: CRC Press, Taylor & Francis Group.

Wiig, S., Storm, M., Aase, K., Gjestsen, M.T., Solheim, M., Harthug, S., Robert, G., Fulop, N. and Team, Q. 2013. Investigating the use of patient involvement and patient experience in quality improvement in Norway: Rhetoric or reality? *BMC Health Serv Res*, 13, 206.

Section IV

Contemporary Nordic Research – Micro-Level Issues

12 Telecare in Home Healthcare Services

Implications for Quality and Safety

Veslemøy Guise, Anne Marie Lunde Husebø, Marianne Storm, Kirsti Lorentsen Moltu and Siri Wiig

CONTENTS

BACKGROUND

Introduction

Home healthcare services are a growing segment of the Norwegian healthcare system. Specialist healthcare services are shifting from the hospital setting to the community (Sibbald et al., 2007; Helse- og Omsorgsdepartementet, 2009), and many healthcare services are increasingly expected to be delivered in people's own homes (Helse- og Omsorgsdepartementet, 2015). A rapidly growing ageing population, many with complex healthcare needs, will necessitate changes to traditional forms of home healthcare provision, such as increased integration of telecare technologies into service delivery (Koch and Hägglund, 2009; Milligan et al., 2011). Telecare solutions can enable healthcare professionals to remotely care for patients living at home and are therefore seen as a promising solution to some of the challenges forecast for future home healthcare services (Solli et al., 2012). Implementation of new information and communication technologies (ICT) in healthcare settings is however complex and is influenced by several factors (Hoonakker, 2012). The attitudes and experiences of healthcare professionals have been flagged as a potential major barrier to telecare implementation (Mair et al., 2007; Brewster et al., 2014; Taylor et al., 2015). A crucial determinant in whether or not healthcare professionals adopt telecare is evidence of its quality and relative advantages to users (Sheikh et al., 2011; Zanaboni and Wootton, 2012). Investigation of healthcare professionals' views on quality and safety may be important, therefore, to help determine the success or otherwise of innovation initiatives like telecare (Brasaite et al., 2014).

Research on the Quality and Safety of Telecare

The study of the quality and safety of telecare is an emerging research field. As a result, knowledge on adverse events and patient safety risks associated with telecare has been lacking and the service quality implications of implementing telecare solutions in home healthcare are poorly understood (Balka et al., 2007; Black et al., 2011; Sheikh et al., 2011; Hoonakker, 2012). Emerging evidence seems to suggest, though, that patient safety risks associated with the use of telecare in home healthcare services may be related to the nature of healthcare tasks, or the characteristics and capabilities of those providing the care, rather than problems with the technology itself (Guise et al., 2014). Evidence on the effectiveness or otherwise of telecare remains inconclusive in terms of both therapeutic effects and increased efficiency in the healthcare services (Ekeland et al., 2010). A number of studies do however report a range of positive outcomes associated with the use of telecare and other ICT-assisted healthcare services for a variety of patient groups (Bowes and McColgan, 2006; Barlow et al., 2007; Ekeland et al., 2010; Lindberg et al., 2013; Husebø and Storm, 2014). There is a need for more research to illuminate the quality and safety implications of using telecare in home healthcare services (Hoonakker, 2012), not least of all studies on implications for patient experiences and patient–provider relationships (Ekeland et al., 2010).

Healthcare Staff Perspectives on the Quality and Safety of Telecare

Staff perspectives on quality and safety in healthcare are contingent on their grasp of professional roles, personal ideas and beliefs (Travaglia et al., 2012; Wiig et al., 2014a). To gain an understanding of how care quality and safety is perceived and enacted in day-to-day provision of telecare services, therefore, we need knowledge of the values, beliefs and professional standards of the healthcare professionals providing the services (Farr and Cressey, 2015). Previous research on staff perspectives on the quality and safety of telecare use has mainly reported on perceived negative implications. For example, some healthcare professionals were found to worry about the adverse consequences of providing care at a distance due to a lack of confidence in the safety and efficiency of telecare (Hibbert et al., 2003). Others were sceptical to whether or not a technology-assisted approach is able to satisfy complex patient needs (Söderlund, 2004), while others again note concern about the impact of technology on therapeutic relationships (Stanberry, 2001). Furthermore, staff may feel that telecare is not the most appropriate or preferred use of their professional skills (Mair et al., 2007, 2008). When telecare is perceived as a threat to care quality and patient safety, for example due to changes in clinical routines, workload and patient interactions, healthcare professionals are likely to resist its implementation and any related changes to the organisation of services (Sharma and Clarke, 2014).

Further understanding of healthcare professionals' perspectives on the quality and safety of care (Wiig et al., 2014a; Farr and Cressey, 2015) could help address potential barriers, risks and resistance to the use of telecare in home healthcare services (Sharma and Clarke 2014; Zhang et al., 2014) and is therefore likely important to aid successful implementation (Hoonakker, 2012¸ Brasaite et al., 2014). The study reported here addresses this need for more knowledge by focusing on the perspectives of healthcare staff on the quality and safety of the planned implementation of virtual home healthcare visits in two Norwegian municipalities. The rationale for using telecare as part of the delivery of primary healthcare services for people living at home is a desire to provide safe, high-quality care while enabling older adults to remain longer in their own homes. Virtual visits, which involve the use of a secure videophone system that enables real-time audio-visual communication between staff and patients, can for example be used to assess a patient's health status, monitor medication routines, demonstrate or supervise procedures and provide social contact (Husebø and Storm, 2014). The study is part of *Safer@Home – Simulation and training*, an action research project where the overall objective is to develop, test and evaluate a training initiative to prepare healthcare providers, patients and their families for taking part in virtual visits (see Wiig et al., 2014b, for a full study protocol).

Aim and Research Question

The aim of this study was to explore healthcare professionals' perspectives on quality and safety implications of implementing virtual home healthcare visits as part of municipal home healthcare services. The following research question has guided the study: *What are healthcare professionals' perspectives on potential quality and safety implications of using virtual visits in the provision of home healthcare services?*

Exploration of the perceptions of healthcare professionals, as key stakeholders in a proposed new telecare service, on potential quality and safety implications of using virtual visits, can contribute to a better understanding of what aspects of care quality and safety that need to be considered prior to implementation, not least in regard to potential training and support needs.

METHODS

Focus group interviews were used to explore healthcare professionals' perspectives on potential quality and safety implications of introducing virtual visits in home healthcare services. The focus group method emphasises group interaction and discussion. It is a particularly apt data collection technique in studies where the aim is to explore participants' attitudes, experiences, beliefs and concerns about an issue, as the group approach taps into a wide variety of frameworks of understanding (Kitzinger, 1995; Kitzinger, 2005). Focus groups have furthermore been recommended when examining professional responses to organisational changes (Kitzinger, 2005), such as the planned implementation of virtual visits. A purposive sampling strategy was used, guided by a desire to include a cross section of health and social care professionals working in the various municipal health and social care services typically delivered in the home context in Norway. Thus, eligible participants were health or social care staff working in different home healthcare or sheltered housing services in the two municipalities planning to use virtual visits as part of service delivery.

Data Collection and Analysis

The data collection was conducted according to the agreed study protocol (Wiig et al., 2014b) during 2013 and 2014. Six focus groups took place in five different types of municipal home health and social care services. Each group consisted of the professional groupings who usually work together in the respective service types, in order to take advantage of homogeneity and existing group dynamics (Krueger and Casey, 2000; Malterud, 2012a). The groups were as follows: three groups with different constellations of registered nurses, enrolled nurses, healthcare workers and care assistants from home healthcare services; one group of physiotherapists from physiotherapy services; one group of occupational therapists from occupational therapist services; and one group of social care workers from sheltered housing services. The same set of topics was covered in each focus group interview: technology experience, implementation of virtual visits in home healthcare services, implications of virtual visits, and training needs. Each of the groups met once, and each interview lasted between 90 and 120 minutes.

The author VG moderated all six focus groups. One or two other members of the research team (SW, MS, AMLH, KM) acted as observers in different focus groups, taking field notes on group dynamics, atmosphere and participant relations (Malterud, 2012a). The degree of participation from the moderator and observers varied across groups. Mainly, the moderator facilitated the group discussion, while the observers contributed with clarifications and additional questions. All interviews were tape-recorded and transcribed verbatim. Data were analysed according to Malterud's (2012b) systematic text condensation approach. All authors contributed to the data analysis. This chapter reports on findings related to 'implications for quality and safety', a code group within the theme 'Implications of using virtual visits'. The study has been ethically approved by the Norwegian Social Science Data Services (Ref. 32934, 16 April 2013). Participation was voluntary, and written informed consent was obtained from all participants. All data have been anonymised and securely stored.

Sample Characteristics

A total of 26 professionals participated in the focus groups. Eighteen were from municipality A (city based, 132.600 inhabitants) and eight from municipality B (rural based, 10.700 inhabitants). The 26 participants included seven registered nurses, four enrolled nurses, five occupational therapists, three physiotherapists, three social workers, one care worker, one social educator, one health worker and one care assistant. There were 23 women and 3 men, aged between 24 and 59 years old (the average age was 39 years). Participants' average length of total working experience was 13.75 years (ranging between 1 and 37 years), with an average 8 years of employment in their respective municipality. Only two participants had previous experience using videophone technology (Skype) at work, but all had experience using technologies such as mobile phones, PCs, digital planning tools and electronic patient records as part of their job.

RESULTS

The analysis revealed that the healthcare professionals perceived both positive and negative quality and safety implications of using virtual visits in home care services. Negative quality and safety implications were a reduced ability for clinical observation and vulnerability of poor assessment of patient needs, while positive quality and safety implications were an increased ability to monitor and follow-up on patients and patient empowerment and reduction of stigma.

Reduced Ability for Clinical Observation

There was broad agreement among the participants that implementation of virtual visits through the use of a video communication tool would diminish health professionals' ability to conduct proper clinical observations of a patient's health status: *we need the clinical gaze!* (Enrolled nurse, Focus Group (FG) 1). Not only nurses, in particular, emphasised this aspect, but also the occupational therapists and physiotherapists expressed the importance of physical presence and ability to conduct physical examinations. The informants considered the potential loss of information gained from clinical observations as the most pressing risk factor associated with virtual visits. Participants' rationale related to the need to observe and directly monitor clinical parameters, for example, by observing skin colour, respiration and ordinary bodily functions, feeling the patients' temperature or noticing uncommon smells from the patient or their surroundings. This is illustrated in the following exchange:

> There is a lot of information you can miss when using such a camera. Because when you visit the patients [at home], you see if they are all right. You don't just observe that the duvet moves up and down, like you would through a camera, but you listen to how they breathe, and you look at their skin colour.
>
> **Registered nurse, FG 1**

> You touch them.
>
> **Enrolled nurse, FG 1**
>
> You feel, you smell, in case they have been sick or have diarrhoea. Yes. So there are a lot of things we check.
>
> **Registered nurse, FG 1**

The participants also described how the physical meeting at the patients' home was key in their observations of not only the patient's health status and bodily functions but also the state of the home environment. One person said: 'In our profession, we are experts in observation. When we are inside a home, we observe everything from ceiling to floor and discover things that we may need to address, and we can lose that of course' (Registered nurse, FG 5). Participants described how the home setting in itself could be a potential risk to patient safety in terms of, for example, lack of cleanliness, poor hygiene, bad air quality and hazards posed by electrical equipment, such as stoves being left on. Healthcare professionals therefore perceived the reduced ability to conduct direct home observations as a potential risk to the safety of virtual visits.

Vulnerability of Poor Assessment of Patient Needs

Moreover, some participants stated that virtual visits would potentially be a more restrictive way of interacting with the patient, with the possible risk of conversations being too focused on the specific topic of the call only and less about emerging topics. Healthcare professionals considered the ability to deal with emerging topics such as patient motivation and deeper personal issues, which often came up during physical home visits, as important information for the quality of the care process. The informants also discussed how virtual visits could have negative implications for relational and communicative aspects of patient care. The informants considered the professional relationship with the patient as key in collecting vital information about the patients' health status and needs. They argued that the virtual visit would give less room for discussions and elaborations and could be a barrier for the patients in sharing information necessary for the healthcare professionals to make the best treatment and care decisions:

> I think the relationship between us and the service user is so important. It is often difficult for them to tell us all they would like to tell us, or express their entire need for help from us. Therefore, the videophone could be a barrier, because it is a bit impersonal. But of course, this depends on the user group. For some users I think this would be great, for others it could be complicated.
>
> **Occupational therapist, FG 3**

Informants emphasised the importance of patient needs assessments prior to implementation of virtual visits, to ensure quality of care. The main issues raised by the participants regarding a needs assessment were related to proper identification of the patient needs that could be addressed via virtual visits without an increased risk to patient safety, the patients' capacity to deal with the new technology, and the patients' ability to learn how to use the technology. There was agreement among participants that these are aspects that need to be properly assessed, to be able to provide sound professional practice via virtual visits: 'My thought is what are the patient's needs? In other words, what are the needs and can we solve everything over the videophone?... I think what the patients' needs are and what tasks need to be solved – those questions should guide the use [of the technology]' (Occupational therapist, FG 3).

Some participants also described worries about whether the use of virtual visits could increase the risk of not getting in touch with patients as scheduled. While it was acknowledged that there could be several reasons for a video call going unanswered, such as the patient simply forgetting the appointment, participants were firm that alternative means of contacting the patient must always be in place, such as making an ordinary phone call or arranging for a prompt physical visit to be carried out:

> If they [the patients] do not answer the [video] call in the first fifteen, twenty minutes, maybe up to an hour, that's one thing. But if we cannot contact them at all, then we'll have to physically drive down [to their home] and see if anything is wrong.
>
> **Social worker, FG 6**

Another safety practice suggested was the need for an error notification system in case of technical problems with either the video communication equipment or the Internet connection. Healthcare professionals stressed that both the user and they themselves must be made aware of any such faults as early as possible, to avoid delays in getting in touch with each other and, not least, in carrying out necessary healthcare procedures.

Furthermore, participants expressed concerns over whether care at a distance could present a potential risk to healthcare professionals' ability to correctly handle patient privacy issues: 'There could be others nearby who could hear [the conversation]. For example, users may have the window open. Or they could have visitors that we do not know of. Which we then do not notice until well into the conversation' (Social worker, FG 6). Such challenges to the staff's duty to preserve patient confidentiality were regarded as a possible threat to overall quality of care.

Increased Ability of Closer Patient Contact and Improved Continuation of Care

The healthcare professionals discussed several positive quality and safety consequences in relation to the use of virtual visits. Participants across all groups agreed that virtual visits would increase opportunities for more frequent patient contact and improved continuation of care: 'You can offer them [patients] more frequent visits and you can offer them increased follow-up with telecare, that I don't doubt' (Physiotherapist, FG 2). Both increased rate and duration of visits were key elements that could improve care quality and safety for a range of patient groups, with a potentially better ability for early detection of changes in a patient's health status, including those with a chronic condition like epilepsy: 'Then we could check on [patients] more often than the single visit we do now. [They] could just as well have a seizure right after we get back into the car' (Registered nurse, FG 1). The same benefits were seen for recently discharged patients: 'Often they need that extra security when they come home after an operation or a short stay [at a nursing home], then they need more frequent visits. So, yes certainly then' (Enrolled nurse, FG 5).

In these ways, virtual visits were seen as a useful supplement to traditional home care visits, allowing for closer patient contact:

> If we only used the videophone it would feel like a poorer and more impersonal interaction. But if you use it [virtual visits] in addition to the ordinary visits, it would be a more comprehensive service and that would increase care quality. You could talk with the patients every day instead of visiting them in their homes only twice a week. You could continue to visit them [physically] one day a week and talk with them every day.
>
> **Occupational therapist, FG 3**

This idea of new telecare tools supplementing traditional home care services to improve its current level of quality and safety was discussed by several participants, with several participants across the groups being adamant that virtual visits must not replace physical visits. Some argued that current services were at a minimum or below acceptable quality level and that healthcare professionals had too little time to perform expected tasks: 'The way things are at present, there is no quality at all … because we are in [the home] for a few seconds, a few minutes maybe and have no time for anything more. What kind of quality is that?' (Enrolled nurse, FG 4). The implementation of virtual visits was thus seen as a way to increase care quality by freeing up time for patients who may need more contact than is currently provided: 'It could ease things a bit for us in relation to how it is today. Then perhaps we could use a little more time on those [patients] who really need it' (Enrolled nurse, FG 4).

While most participants felt that virtual visits could improve the quality and safety of home healthcare services by providing increased opportunities for more frequent patient contact and better continuity of care, the idea of using telecare tools was quite contentious within one of the focus groups, as is illustrated by the following exchange:

> I think it is terrible, because they [municipal decision-makers] don't consider that the most important thing is after all the person [healthcare professional] who physically walks through the patient's door every day to see if they are ok. That is much more important than all these robots and everything. I think it is absolutely awful. They don't think about the social aspect, they don't think about the human aspect at all.
>
> **Care assistant, FG 4**

> But we can't stop it [the use of telecare], you know. That's how service development is.
>
> **Enrolled nurse, FG 4**

> Yes, but it's terrible either way, to think about that development, because these are people who are stuck in a chair, perhaps all day.
>
> **Care assistant, FG 4**

> How do we treat them [patients] today, though? We are there [at their home] for two seconds.
>
> **Enrolled nurse, FG 4**

This discussion clearly shows that while healthcare professionals in general are very concerned about the safety and quality of the healthcare services they provide, there are rather differing understandings of how best to achieve safe, good quality care. While many agreed that the quality and safety of home healthcare could ultimately be improved by the introduction of virtual visits, not all participants were convinced.

Patient Empowerment and Reduction of Stigma

One quality improvement aspect of implementing virtual visits mentioned by the participants was the potential for improved patient empowerment. Some felt that virtual visits could be a good way for healthcare professionals to give more responsibility for own care over to patients: 'That's what's really our problem, we so easily go in and do things for them, so that's almost the hardest. And in these cases it may be that this [virtual visits] could be good' (Enrolled nurse, FG 5). Some participants also mentioned how the implementation of virtual visits could improve care quality by encouraging more self-care practices when patients return home from a healthcare institution, a time when they are often vulnerable and need close follow-up: 'In this phase we are experts in making them helpless. We do everything for them' (Nurse, FG 5). 'Yes, that's when we could really make use of this [virtual visits] in a positive way' (Enrolled nurse, FG 5).

In addition, quality improvement implications were conceptualised as the possibility of being able to reduce real or imagined stigma of being a care recipient and having the home care services visibly visiting one's home several times a day: 'It [a virtual visit] is not visible [to others], so then you can [help the patient] in a way that implies the least amount of visibility to others' (Occupational therapist, FG 3). The use of virtual visits was here associated with a less publicly visible approach to home healthcare provision and was seen as a way of better preserving patient privacy and enabling services more tailored to patients' individual preferences, potentially improving overall care experiences.

DISCUSSION

This focus group study explored healthcare professionals' perspectives on quality and safety implications of using virtual visits in home healthcare services. The findings show that healthcare professionals perceive negative as well as positive consequences for the quality and safety of the care they provide to patients living at home. Staff perceptions of both positive and negative implications were found to have close links to individual views of ideal and appropriate clinical roles and task performances, as well as their day-to-day experiences concerning patient interactions and service organisation (Travaglia et al., 2012; Farr and Cressey, 2015). There was also some degree of disagreement among participants on the potential quality and safety implications of virtual visits, which were mainly reflected in rather contradictory perceptions regarding, on the one hand, a diminished ability to observe and monitor patients and, on the other hand, an increase in patient contact and continuity of care.

The main risks to patient safety that participants foresaw were a reduced ability to sufficiently observe patients' health statuses and monitor clinical parameters and any potential hazards in the home environment. In addition, they were concerned about the potential risk of not being able to adequately detect and assess patients' needs during virtual visits. These findings are in line with previous research addressing real or perceived risks to home healthcare quality and safety associated with the use of telecare in the Nordic countries (Söderlund, 2004; Nilsson et al., 2008; Wälivaara et al., 2011; Reierson et al., 2015) and elsewhere (Hibbert et al, 2004; Mair et al., 2007; 2008; Brewster et al., 2014; Guise et al., 2014; Sharma and Clarke, 2014).

At the crux of these established concerns about negative implications for care quality and patient safety are apprehensions about the loss of conventional means of knowing and caring for patients (Nagel et al., 2013). 'Knowing the patient' is recognised as a core component of safe, high-quality healthcare (Luker et al., 2000; Zolnierek, 2014). Traditionally, knowing happens through communication and use of the senses by way of physical presence (Bundgaard et al., 2012), as was described by our participants. Since the use of telecare implies physical distance between provider and patient, it suggests fundamental changes to patient–provider interactions, which for many healthcare professionals do not fit with their view of how to provide safe and effective care (Sharma and Clarke, 2014). One of the main challenges of sustained telecare usage are current gaps in the knowledge of how to 'know the patient' in virtual environments (Nagel et al., 2013). As was suggested by our participants, thorough and ongoing patient needs assessments both before and during the use of telecare services can be a way to ensure that patients' suitability for using these types of services are regularly monitored and assessed.

Our study also found that healthcare professionals foresee several positive quality and safety implications of using virtual visits in home healthcare services. Conceptualisations of positive quality implications are interesting as this is something that, in contrast to negative quality and safety implications, has been given little attention in the previous literature. Some of these findings can be seen as a reflection of ongoing Norwegian healthcare policies and reforms, mainly the move towards providing more care in community settings (Helse- og Omsorgsdepartementet, 2009, 2015), thus giving municipal primary care services increased responsibility for growing numbers of vulnerable patients. Study participants believed that virtual visits could facilitate increased frequency and duration of patient contact, in part by freeing up more time for longer face-to-face visits than what current service organisation seemingly allows for. Telecare was thus seen to have the potential to improve the quality and safety of home healthcare by increasing accessibility to professional care and offering more opportunities for enhanced patient follow-up and monitoring where needed (Nilsson et al., 2010; Sharma and Clarke 2014; Kajander and Storm, in press). As was noted by some participants, this could be especially valuable for vulnerable patients newly discharged from hospital (Hesselink et al., 2012; Laugaland et al., 2012; Storm et al., 2014). Virtual visits have indeed been found to facilitate continuous and coordinated care (Husebø and Storm, 2014), as well as quick patient access to qualified healthcare staff (Kajander and Storm, in press) during periods of transition from hospital to home-based care.

Another key finding regarding positive implications for care quality was that virtual visits were seen to enable and support increased patient involvement in own care. Again, the findings here could be explained by recent government policies emphasising a substantial strengthening of patient participation and empowerment within the Norwegian healthcare system as a whole (Helse- og Omsorgsdepartementet, 2013, 2014). Study participants admitted that rather than encouraging patients' independence and self-care, current practice sometimes contributed to helplessness and dependency. They therefore saw the potential for increasing and upholding patient participation as a fundamental advantage of virtual visits. Certainly, telecare solutions can offer a flexible and efficient way for professionals to provide advice and guidance in support of a variety of self-care practices in home-dwelling patients with long-term conditions (Bond, 2014; Husebø and Storm, 2014; Kajander and Storm, in press). The fact that new ICT solutions can have a positive impact on patients' autonomy and independency has furthermore been noted as a crucial incentive to their being adopted for use by healthcare professionals in the home healthcare services in Norway (Gjestsen et al., 2014).

Few participants had practical experience of using either telecare in general or videophone technology for virtual visits specifically. They were thus speaking only from the perspective of potential future users of such telecare tools and not as experienced telecare professionals. The findings reported here should therefore be viewed in this context, although they do mirror outcomes from a Swedish study where nurses experienced that the use of telecare increased the quality of home healthcare (Nilsson et al., 2010). It is also important to note that participants' perspectives on the implications of proposed virtual visits varied, as has been found elsewhere (Taylor et al., 2015). Some argued quite strongly that virtual visits did not fit with their ideals of good quality in healthcare service provision and would be impersonal, dehumanising and increase the distance between the service provider and the patient. Mair et al. (2008) note similar findings. Others again were very encouraging and could see a lot of potential for quality improvement through virtual visits. Many were eager to try out this new way of working and communicating with patients, as long as it was used with the right group of patients and if staff were given sufficient information and training beforehand (Zhang et al., 2014; Taylor et al., 2015).

CONCLUSIONS

To ensure safe, purposeful and effective use of telecare services and to aid long-term implementation and adoption of telecare tools in municipal healthcare settings, it is important to have knowledge and understanding of healthcare professionals' views on implications for care quality and patient safety, not least in regard to potential training and support needs. In this study, staff perceptions of possible quality and safety implications of using virtual visits in home healthcare were explored. While findings concerning perceived negative implications for quality and safety are similar to those found in previous research independent of geographical location and context, the positive quality and safety implications observed here could be related to current Norwegian healthcare policy. Perceptions of both positive and negative implications were found to be linked to views on necessary and appropriate clinical competencies and ideal patient–provider relationships in dynamic domestic settings, all within the structures imposed by a complex municipal healthcare system. This study indicates then how multiple contextual factors influence healthcare professionals' conceptualisations of the quality and safety implications of using telecare in the provision of home healthcare services.

ACKNOWLEDGEMENTS

The Safer@Home – simulation and training study is part of the project 'Smart systems to support safer independent living and social interaction for elderly at home' (Safer@Home). This project is supported by the Norwegian Research Council grant number 210799. The authors would like to thank the focus group participants for their kind contribution. They would also like to thank the funders and the Department of Health Studies, University of Stavanger, for the opportunity to carry out this research. The authors would also like to thank partners in the overall Safer@Home project: Lyse; Department of Electronical Engineering and Computer Sciences at University of Stavanger; Stavanger municipality; Cisco; DevoTeam; VS-Safety; SINTEF; SAFER/Laerdal Medical and Stavanger University hospital/SESAM.

REFERENCES

Balka, E., Doyle-Waters, M., Lecznarowicz, D. and FitzGerald, J.M., 2007. Technology, governance and patient safety: Systems issues in technology and patient safety. *International Journal of Medical Informatics*, 76, 35–47.

Barlow, J., Singh, D., Bayer, S. and Curry, R., 2007. A systematic review of the benefits of home telecare for frail elderly people and those with long-term conditions. *Journal of Telemedicine and Telecare*, 13(4), 172–179.

Black, A.D., Car, J., Pagliari, C., Anandan, C., Cresswell, K., Bokun, T., McKinstry, B., Procter, R., Majeed, A. and Sheikh, A., 2011. The impact of ehealth on the quality and safety of health care: A systematic overview. *PLoS Medicine*, 8, 1–16.

Bond, C.S., 2014. Telehealth as a tool for independent self-management by people living with long term conditions. *Studies in Health Technology and Informatics*, 206, 1.

Bowes, A. and McColgan, G., 2006. *Smart Technology and Community Care for Older People: Innovation in West Lothian, Scotland*. Edinburgh, Scotland: Age Concern Scotland.

Brasaite, I., Kaunonen, M. and Suomien, T., 2014. Healthcare professionals' knowledge, attitudes and skills regarding patient safety: A systematic literature review. *Scandinavian Journal of Caring Sciences*, 29(1), 30–50.

Brewster, L., Mountain, G., Wessels, B., Kelly, C. and Hawley, M., 2014. Factors affecting front line staff acceptance of telehealth technologies: A mixed-method systematic review. *Journal of Advanced Nursing*, 70(1), 21–33.

Bundgaard, K., Nielsen, K.B., Delmar, C. and Sørensen, E.E., 2012. What to know and how to get to know? A fieldwork study outlining the understanding of knowing the patient in facilities for short-term stay. *Journal of Advanced Nursing*, 68(10), 2280–2288.

Ekeland, A.G., Bowes, A. and Flottorp, S., 2010. Effectiveness of telemedicine: A systematic review of reviews. *International Journal of Medical Informatics*, 79(11), 736–771.

Farr, M. and Cressey, P. 2015. Understanding staff perspectives of quality in practice in healthcare. *BMC Health Services Research*, 15(1), 123.

Gjestsen, M.T., Testad, I. and Wiig, S., 2014. What does it take? Healthcare professionals perspective on incentives and obstacles related to implementing ICTs in home-based care for the elderly. *BMC Health Services Research*, 14(Suppl. 2), 45.

Guise, V., Anderson, J. and Wiig, S., 2014. Patient safety risks associated with telecare: A systematic review and narrative synthesis of the literature. *BMC Health Services Research,* 14, 588.

Helse- og Omsorgsdepartementet, 2009. Meld. St. 47 (2008–2009). *Samhandlingsreformen – Rett behandling, på rett sted, til rett tid.* (White paper). Oslo, Norway: Det Kongelige Helse- og Omsorgsdepartement.

Helse- og Omsorgsdepartementet, 2013. Meld. St. 29 (2012–2013) *Morgendagens Omsorg.* (White paper). Oslo, Norway: Det Kongelige Helse- og Omsorgsdepartement.

Helse- og Omsorgsdepartementet, 2014. Meld. St. 11 (2014–2015) *Kvalitet og pasientsikkerhet 2013.* (White paper). Oslo, Norway: Det Kongelige Helse- og Omsorgsdepartement.

Helse- og Omsorgsdepartementet, 2015. Meld. St 26 (2014–2015). *Fremtidens primærhelsetjeneste – nærhet og helhet.* (White paper). Oslo, Norway: Det Kongelige Helse- og Omsorgsdepartement.

Hesselink, G., Schoonhoven, L., Barach, P., Spijker, A., Gademan, P., Kalkman, C., Liefers, J., Vernooij-Dassen, M. and Wollersheim, H., 2012. Improving patient handovers from hospital to primary care. A systematic review. *Annals of Internal Medicine*, 157, 417–428.

Hibbert, D., Mair, F.S., Angus, R.M., May, C., Boland, A., Haycox, A., Roberts, C., Shiels, C. and Capewell, S., 2003. Lessons from the implementation of a home telecare service. *Journal of Telemedicine and Telecare*, 9(Suppl. 1), 5–6.

Hibbert, D., Mair, F.S., May, C.R., Boland, A., O'Connor, J., Capewell, S. and Angus, R.M., 2004. Health professionals' responses to the introduction of a home telehealth service. *Journal of Telemedicine and Telecare*, 10(4), 226–230.

Hoonakker, P., 2012. Human factors in telemedicine. In: P. Carayon, ed., *Human Factors and Ergonomics in Health Care and Patient Safety* (2nd edn.). Boca Raton, FL: CRC Press, Taylor & Francis Group, pp. 293–305.

Husebø, A.M.L. and Storm, M., 2014. Virtual visits in home health care for older adults. *The Scientific World Journal*, [online] Available at: http://dx.doi.org/10.1155/2014/689873 (accessed 18 June 2015).

Kajander, M. and Storm, M., In press. 'Kontakt med ett trykk': Videotelefonsamtaler mellom hjemmeboende brukere og helsepersonell. *Nordisk Sygeplejeforskning*. In press. [In Norwegian].

Kitzinger, J., 1995. Qualitative research: Introducing focus groups. *British Medical Journal*, 311(7000), 299–302.

Kitzinger, J., 2005. Focus group research: Using group dynamics. In: I. Holloway, ed. *Qualitative Research in Health Care.* U.K., England: Open University Press, pp. 56–70.

Koch, S. and Hägglund, M., 2009. Health informatics and the delivery of care to older people. *Maturitas*, 63, 195–199.

Krueger, R.A. and Casey, M.A., 2000. *Focus Groups: A Practical Guide for Applied Research*, 3rd ed. Thousand Oaks, CA: Sage Publications.

Laugaland, K., Aase, K. and Barach, P., 2012. Interventions to improve patient safety in transitional care – A review of the evidence. *Work*, 41, 2915–2924.

Lindberg, B., Nilsson, C., Zotterman, D., Söderberg, S. and Skär, L., 2013. Using information and communication technology in home care for communication between patients, family members, and healthcare professionals: A systematic review. *International Journal of Telemedicine and Applications*, 2013, 2.

Luker, K.A., Austin, L., Caress, A. and Hallett, C.E., 2000. The importance of 'knowing the patient': Community nurses' constructions of quality in providing palliative care. *Journal of Advanced Nursing,* 31(4), 775–782.

Mair, F., Finch, T., May, C., Hiscock, J., Beaton, S., Goldstein, P. and McQuillan, S., 2007. Perceptions of risk as a barrier to the use of telemedicine. *Journal of Telemedicine and Telecare*, 13(1), 38–39.

Mair, F.S., Hiscock, J. and Beaton, S.C., 2008. Understanding factors that inhibit or promote the utilization of telecare in chronic lung disease. *Chronic Illness*, 4(2), 110–117.

Malterud, K., 2012a. *Fokusgrupper som forskningsmetode for medisin og helsefag*. Oslo, Norway: Universitetsforlaget. [In Norwegian].

Malterud, K. 2012b. Systematic text condensation: A strategy for qualitative analysis. *Scandinavian Journal of Public Health*, 40(8), 795–805.

Milligan, C., Roberts, C. and Mort, M., 2011. Telecare and older people: Who cares where? *Social Science and Medicine*, 72, 347–354.

Nagel, D.A., Pomerleau, S.G. and Penner, J.L., 2013. Knowing, caring, and telehealth technology: "Going the distance" in nursing practice. *Journal of Holistic Nursing*, 31, 104–112.

Nilsson, C., Skär, L. and Söderberg, S., 2008. Swedish district nurses' attitudes to implement information and communication technology in home nursing. *The Open Nursing Journal*, 2, 68–72.

Nilsson, C., Skär, L. and Söderberg, S., 2010. Swedish District Nurses' experiences on the use of information and communication technology for supporting people with serious chronic illness living at home – A case study. *Scandinavian Journal of Caring Sciences*, 24(2), 259–265.

Reierson, I.Å., Solli, H. and Bjørk, T.I., 2015. Nursing students' perspectives on telenursing in patient care after simulation. *Clinical Simulation in Nursing*, 11, 244–250.

Sharma, U. and Clarke, M., 2014. Nurses' and community support workers' experience of telehealth: A longitudinal case study. *BMC Health Services Research*, 14, 164.

Sheikh, A., McLean, S., Cresswell, K., Pagliari, C., Pappas, Y., Car, J., Black, A. et al. 2011. *The Impact of eHealth on the Quality and Safety of Healthcare: An Updated Systematic Overview and Synthesis of the Literature.* Final report for the NHS Connecting for Health Evaluation Programme. Edinburgh, Scotland: The University of Edinburgh.

Sibbald, B., McDonald, R. and Roland, M., 2007. Shifting care from hospitals to the community: A review of the evidence on quality and efficiency. *Journal of Health Services Research & Policy*, 12(2), 110–117.

Söderlund, R., 2004. The role of information and communication technology in home services: Telecare does not satisfy the needs of the elderly. *Health Informatics Journal*, 10, 127–137.

Solli, H., Bjørk, I. T., Hvalvik, S. and Hellesø, R., 2012. Principle-based analysis of the concept of telecare. *Journal of Advanced Nursing*, 68(12), 2802–2815.

Stanberry, B., 2001. Legal ethical and risk issues in telemedicine. *Computer Methods and Programs in Biomedicine*, 64, 225–233.

Storm, M., Siemsen, I.M.D., Laugaland, K., Dyrstad, D.N. and Aase, K., 2014. Quality in transitional care of the elderly: Key challenges and relevant improvement measures, *International Journal of Integrated Care*, 14(8), 1–15.

Taylor, J., Coates, E., Brewster, L., Mountain, G., Wessels, B. and Hawley, M.S., 2015. Examining the use of telehealth in community nursing: identifying the factors affecting frontline staff acceptance and telehealth adoption. *Journal of Advanced Nursing*, 71(2), 326–337.

Travaglia, J.F., Nugus, P.I., Greenfield, D., Westbrook, J.I. and Braithwaite, J., 2012. Visualising differences in professionals' perspectives on quality and safety. *BMJ Quality and Safety*, 21, 778–783.

Wälivaara, B.M., Andersson, S. and Axelsson, K., 2011. General practitioners' reasoning about using mobile distance-spanning technology in home care and in nursing home care. *Scandinavian Journal of Caring Sciences*, 25(1), 117–125.

Wiig, S., Aase, K., von Plessen, C., Burnett, S., Nunes, F., Weggelaar, A.M., Anderson-Gare, B., Calltorp, J. and Fulop, N., for QUASER-team, 2014a. Talking about quality: Exploring how 'quality' is conceptualized in European hospitals and healthcare systems. *BMC Health Services Research*, 14, 478.

Wiig, S., Guise, V., Anderson, J., Storm, M., Husebø, A.M.L., Testad, I., Søyland, E. and Moltu, K., 2014b. Safer@Home – simulation & training: The study protocol of a qualitative action research design. *BMJ Open* 4, e004995 [online]. Available at: http://bmjopen.bmj.com/content/4/7/e004995.full (accessed 17 August 2015).

Zanaboni, P. and Wootton, R., 2012. Adoption of telemedicine: From pilot stage to routine delivery. *BMC Medical Informatics and Decision Making*, 12, 1.

Zhang, W., Barriball, K.L. and While, A.E., 2014. Nurses' attitudes towards medical devices in healthcare delivery: A systematic review. *Journal of Clinical Nursing*, 23(19–20), 2725–2739.

Zolnierek, C.D., 2014. An integrative review of knowing the patient. *Journal of Nursing Scholarship*, 46(1), 3–10.

13 Coping with Complexity

Sensemaking in Specialised Home Care

Mirjam Ekstedt

CONTENTS

INTRODUCTION

Healthcare systems face considerable challenges in the coming years as a consequence of current demographic changes: a growing and ageing population and new groups of patients with chronic and complex disorders. Concurrent medical and technical advances have made it possible to move the administration of potent drugs and the use of complex medical technology into patients' homes (Fex et al., 2009). The rapid medical advances increase professional specialisation, making multiple competencies imperative in assessments and decisions regarding a patient's care. Optimising safety and continuity of care in the face of these challenges imposes great demands on healthcare organisations and requires a radical rethinking of ways to support patients in self-management at home.

In order to understand how safety is created in specialised home care settings, it is necessary to know more about the cognitive strategies used by professionals at the sharp end of practice, to manage problem-solving and decision-making in everyday clinical work (ECW) (Klein, 1998). This chapter aims to contribute to this understanding through the lens of sensemaking theory (Weick, 1995; Weick et al., 2005).

COMPLEXITY IN SPECIALISED HOME CARE

Specialised home care in Sweden is provided in a patient's private home on a 24-hour basis by an ambulatory multidisciplinary care team, as an alternative to hospital care. Although most of these specialised home care teams serve patients with cancer, the volume and scope of such practices are expected to expand over the next decade since it will also include patients with multi-morbidity and complex chronic conditions. The complexity of specialised home care is qualitatively different from that delivered in institutional settings (Ekstedt and Cook, 2014). The range of patient illnesses and conditions that a caregiver might encounter is greater than in most institutional settings. Many specialised professionals from different care providers, governed by different values and laws (the Social Services Act and the Health Care Act), are usually involved in the care of a single patient. This, combined with the underlying patient complexity, makes home care ripe for developing 'gaps' in communication and information transfer (Caines et al., 2011; Abraham et al., 2012).

Characteristics of specialised home care organisations are consistent with features of complex adaptive systems (CASs) theory (Ciliers, 1998). CASs are open systems, meaning that it is difficult to frame the boundaries surrounding the system and that the borders of responsibility, for example, may be blurred. In a complex system, each component is ignorant of the behaviour of the system as a whole and does not even know the full effects of its own actions. This means that the interactive relationships surrounding a patient's care trajectory are dynamic, non-linear, difficult to foresee and more or less impossible to control (Dekker, 2011). Features of CAS are adaptation and self-organisation, in the sense that personnel constantly adjust to the obstacles and messy details that occur in ECW. CAS also has a memory, in the sense that history – each individual's and organisation's former experiences of what works or not – co-determines self-organisation. Relationships are a key feature of CAS, which leads us to consider sensemaking as a phenomenon emerging from iterative reciprocal interactions between agents involved in care.

At the operative point of specialised home care, where decisions and actions affect patients directly, the physician in charge of the patient and the multidisciplinary team are generally responsible for planning care and setting goals. Nevertheless, the work is for the most part conducted by single professionals going individually to a patient's home to provide care. Decisions about changes in treatment or other actions rely on direct observations, patient conversations and physical examinations made during these regular visits. Laboratory studies require re-planning and coordination of care, and imaging services will involve moving the patient to an imaging facility. Thus, single professionals in the multidisciplinary team make countless rapid-fire decisions on a daily basis, which have an impact on the patients as well as on the staff's conditions for care. A previous study documents how practitioners in home care effectively use tailored responses to uncertainty and disturbances in thoughtful, inventive, courageous, meaningful, clever, inspired, determined and purposeful ways (Ekstedt and Cook, 2014). However, there is a paucity of empirical support regarding how professionals make sense of, and cope with, the complexity in home care to reach specific goals. The same is true of how the responses to unexpected situations are transformed into learning and get incorporated into habitual work.

The theory of sensemaking provides a useful conceptual link between the individual and organisational processes of decision-making in this context (Hayes, 2013). This perspective allows both individual aspects and organisational circumstances to be integrated into the individual decisions about a road of action. In the following sections, the theoretical bases of sensemaking will first be described briefly and then adopted into a tentative framework that will be used for illustrating the empirical study on ECW in specialised home care.

THEORETICAL PERSPECTIVE ON SENSEMAKING

Clinical cognitive work is devoted to making and sustaining a coherent, purposeful and relevant understanding of patients and their conditions. Weick (1995) describes this as *sensemaking. Sensemaking* is about the interplay of actions and interpretations, which helps individuals or a collective of individuals to build consensus around a situation and provide a road of action by sharing their unique perspectives.

Sensemaking has diverse theoretical routes and has been explored in a wide variety of domains and disciplines: organisational (Weick et al., 2005), educational (Russell et al., 1993; Warren et al., 2001; Jordan et al., 2009) and human–computer interaction (Savolainen, 1993; Nosek, 2005), to mention just a few. Sensemaking has also been viewed as a more individual activity. For example, Rings and Rands defined sensemaking as 'a process in which individuals develop cognitive maps of their environment' (1989, p. 342). More recently, sensemaking has been adopted in the endeavour to enhance self-management strategies in chronic diseases (Mamykina et al., 2015). In this view, *understanding* is the term that describes mutual activity.

Sensemaking is *grounded in identity constructions*, deriving from the need to organise flux when the flow of routine actions is interrupted. Sensemaking starts with noticing and bracketing: individuals examine the situation at hand and try to classify it in relation to their existing mental models of related phenomena. Sensemaking is also about *labelling*: individuals give shape to their lived experiences through verbal descriptions, which allow them to share their meanings with one another, thereby contributing to the development of common ground. Sensemaking is *retrospective* in the sense that individuals construct meaning out of situations by 'stepping outside the stream of experience and direct attention to it' (Schutz, 1967). This can only be done after they have completed their involvement and can reflect on the outcomes. Sensemaking is about *presumption* and the most plausible hypothesis. Sensemaking is *social and systemic*: an individual's sensemaking is shaped by, and in turn shapes, the opinions of others. Considering the modest amount of empirical work on sensemaking that has been accumulated so far, this study in out-of-hospital care will fill a gap in understanding of the largely invisible social processes which not only are taken for granted but also woven into communication and activities in specialised home care.

In later works, Weick and his fellow researchers state that sensemaking is a process that is 'ongoing, social and easily taken for granted' (2005, p. 409). The inference of this statement is that humans under ordinary conditions adapt to a habitual mode of sensemaking, which is so natural that it is not even thought about. He goes on to state that explicit efforts at sensemaking tend to occur 'when the current state of the world is perceived to be different from the expected state of the world, or when there is no obvious way to engage the world' (Weick et al., 2005, p. 409). Sensemaking means that 'the interplay of action and [the fact] that interpretation, rather than the influence of evaluation of choice, is the core phenomenon' (Weick et al., 2005, p. 409). These statements suggest that when gaps of understanding occur, they are triggering explicit efforts to make sense of the uncertain or ambiguous. This is a contrast to the *ongoing habitual mode* of sensemaking. Further, they suggest that this *explicit mode of sensemaking* is characterised by an interplay of *action* and *interpretation*, whether by making inferences from earlier experiences through an inner communication or by interaction and communication with others. These cornerstones are adopted into the tentative model of sensemaking in ECW, further described overleaf.

Tentative Model for Sensemaking in Everyday Clinical Work

The tentative model for sensemaking in ECW presented in the following (Figure 13.1) is based on the theoretical perspectives of Weick (1995) and Weick et al. (2005) and inspired by the framework in diabetes self-management created by Mamykina and co-workers (2015). This model will be used to illustrate and discuss the empirical findings from specialised home care in this study.

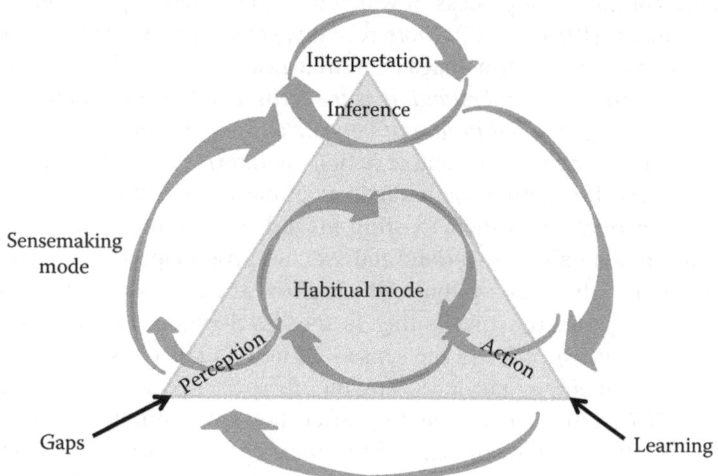

FIGURE 13.1 Tentative model of sensemaking in everyday clinical work. Inner circle, *habitual mode*, a 'default state' characterised of routine actions, implicit understanding and use of existing mental models; outer circle, *sensemaking mode*, characterised of explicit, mindful and active use of cognitive skills; left corner of the triangle, *perception*, monitoring and classification of new information and experiences; *upper corner*, construction of new *Inferences* allow for a certain road of action and *interpretation* of plausible models in search for an answer to the question: 'what is the story?'; *right corner*, the process of carrying out ECW in response to new information; *perception of gaps* in understanding disrupts the flow of action and initiates sensemaking mode; *learning*, new understanding is integrated as routine in the habitual mode.

The proposed model suggests that sensemaking in ECW is carried out in one of two modes: *habitual mode* (inner circle Figure 13.1) and *sensemaking mode* (outer circle Figure 13.1). The model also proposes that both the habitual mode and the sensemaking mode involve three essential activities (represented by the three corners in the triangle): (1) *perception*, which involves the monitoring and noticing of new pieces of information and cues of situations in ECW; (2) *inference*, which implies the activation of relevant patterns of knowledge and experiences in an active interplay of action and *interpretation* of mental models in search of the most plausible road of action (inferences and interpretations are made in an inner communication and/or in interaction with others); and (3) *action*, which is the process of carrying out ECW in response to new information.

The *habitual mode* is characterised by implicit understanding, use of existing rules and routines, containing an anticipated flow of actions. Professionals interpret and respond to the perception of new information so naturally that they do not even think of this as an effort. In the habitual mode, the *perception–inference–action* cycle activates existing experiences and mental models rather than creating a gap in understanding, and professionals fall into routine actions rather than choosing a new road of actions.

The *sensemaking mode* is characterised by an explicit, mindful and effortful use of cognitive skills in which professionals analytically engage and interact with the new information or situation at hand. Explicit efforts at sensemaking are triggered when *gaps* in understanding are identified and when the flow of routine actions is interrupted (left corner). In sensemaking mode, interpretation and inference of plausible models interplay in the search for an answer to the question: 'what is the story here?' (top corner). Explicit efforts are made to make the world comprehensive in order to find the best road of action (right corner). The new understanding that emerges becomes integrated as routine in the habitual mode and may reshape routines and lead to organisational learning.

EMPIRICAL STUDY

The chapter is based on an empirical study, conducted over 2 years in three specialised (palliative) home care units in the county of Stockholm. The three care locations were selected in order to increase the geographic and sociodemographic range of settings and internal routines. The study, which was carried out at the 'sharp end' of care, that is, at the point of contact between the patient and the medical system, explores what professionals actually *do* to deliver safe and coordinated home care to patients with complex disorders (Woods, 1993). Specific foci were sensemaking and communication between members in the multidisciplinary home care teams under ordinary conditions. The research was approved by the appropriate institutional review board for human subjects. Methods and settings are briefly described in the following. For a thorough description, see Ekstedt and Cook (2014).

METHODS

We applied an ethnological perspective using several observational techniques, in order to acquire knowledge 'from within' to see how problems were framed and understood (Schutz, 1970). The methods, successfully used in other fields (Woods, 1993; Albolino et al., 2007; Arman et al., 2012), provided reliable data on how work was actually performed. The observational methods included general participant observations of 'places' of care, and 'shadowing' professionals in ECW, using a semi-structured protocol, field notes, audio recordings and photographs. The research team also performed short interviews with key actors (including patients and family caregivers) to clarify and deepen the observations. During the field observations, we participated in formal morning meetings, handover situations and during phone calls and chitchat throughout the day where team members prepared for home care encounters, shared information and updated their knowledge during the shift. We also followed clinicians during their daily home care encounters in order to understand the complexity of work done and see how they made sense of situations and stored and communicated key information. We paid particular attention to situations where the work tasks prompted patient-centred or inter-worker communications. Interviews and observations were tape-recorded and transcribed verbatim. Traces of sensemaking and actions taken were analysed by the use of thematic analysis and process-tracing methodologies described by Woods (1993). The goal in these methods is to map out how an incident unfolded, including available cues, in both the immediate and the larger context. The observations have been integrated with interview and field notes. The main findings from the analysis on sensemaking in ECW are presented and illustrated with quotations and cases.

Preparation: Putting Together the Pieces of the Puzzle

ECW in home care is exposed to an infinite stream of events and inputs that surround the professionals. Decisions rely on each expert's pre-understanding of a patient's most recent status and likely trajectory. Ordinary clinical work is made up of periods of attention to different tasks. Interruptions are common, and each interruption requires a shift of attention to a different focal point. Every shift of attention from a patient or task to an interrupting issue and back again requires a process of understanding. So common is this activity that it goes almost completely unnoticed by clinicians themselves. Conversations that went on during the preparations for the day were characterised by isolated sentences or short calls that could not be understood by an outsider, but for those involved made up the pieces of the puzzle needed to perform ECW.

Case 13.1 Morning Preparation

Before the morning meeting: A team of nurses and assistant nurses prepare for the day ahead by reading medical charts on the computer, printing to do-lists and talking in smaller groups. Random sentences are overheard:

- Automatic infusion pump, but it doesn't say what it is, what should be in it … It just says '50 mg furo' [i.e. *furosemide, diuretic*] …

The phone rings and a nurse answers. Her side of the conversation is as follows:

- The doctor said that everything connected to x-rays, you should contact the oncologist … they are the ones that are supposed to check up on you and they should write the referral.
- I'll leave a message for M/the doctor. She'll be back next week, so she'll know there's been a problem.
- Okay, well, I'll pass that along …

During the morning meeting: The doctor asks for feedback on a prescription given earlier. The nurse, who knows the patient, explains and gives a few small comments on what should be observed when they visit the patient.

The doctor: I haven't seen the results of the tests that were supposed to be done on XX yesterday.
The nurse: Yeah, but we were wondering if they could be done on Friday, when we have a visit there.
The doctor: NO, this cannot wait until then. CREA [i.e. *Serum-creatinine, an important indicator of renal health*] was very high and we have to do the test today, there's a risk of complications in the kidneys.

Field note: It turned out that one nurse had never seen the dialysis equipment that a patient had at home. Various people tried to explain how it worked. The other nurses tried to explain the various connections, and the end result is that staff 'traded' patients with each other.

Short interview after the morning meeting:
The morning meetings can seem chaotic, but I think it's great that we go over everything. It's enough to say the name of someone you're going to see to jog your memory and if you have any questions and … Just saying the names is enough – it can mean many things, for instance if the assistant nurse is seeing the same patient and sometimes we can coordinate – for instance if I'm just delivering medication and they are going anyway, they can do that, so I don't have to rush. And then … You can't remember everything, maybe you haven't written everything down … Then you can say: 'don't forget this and remember that'. It's hard to write everything down …

This case illustrates how 'order' is constructed through the use of spoken and written 'words'. By naming patients, prescriptions or work tasks aloud, an interactive process of sensemaking started in a variety of ways that had real consequences for the choice of actions. Decisions were made for structuring workflow, avoiding mistakes and creating continuity of care.

The morning *preparation* case also showed how new or deficient information created gaps in understanding, triggering efforts at sensemaking. The perception of the missing lab revealed brittleness in routines and highlighted gaps in the transfer of information and a discrepancy between the nurse's and the physician's view about the urgency of the test. In habitual work, nurses coordinate routine blood tests with their ordinary patient visits. However, in this instance, the nurse was not aware that the test should have led to an extra visit to the patient, resulting in a discussion and interactive interplay between professionals (sensemaking mode).

Building Understanding at a Distance

Although specialised home care is surrounded by a multidisciplinary team, the observations revealed that ECW to a large extent is performed in 'professional silos'. We observed how single professionals built their understanding of a patient in their first few minutes of a home care encounter. During the short encounter, the nurse notices changes in the patient's status, makes inferences in relation to what is expected and tries to identify vital signs of change for closer attention. During this initial state of sensemaking, the nurse tries to get a sense of: *what is going on here*? The following case illustrates how the nurse becomes aware of vital signs that are at variance with the person's 'normal' conditions. Attaining this initial understanding is critical because it sets the stage for everything that happens afterwards. The following example is from an interview with nurses.

Case 13.2 Sensemaking at a Distance

Nurse 1: No, but all these assessments where you feel pretty much alone ... this is mostly about us capturing stuff.

Interviewer: How do you do that?

Nurse 4: Well, I try to keep in mind that it's my job to do a full workup in so far as possible. If the patient is in pain, we'll discuss that for a while to find out more. What kind of pain is it, where and so on ...

Nurse 2: Then I'll report all my observations to a doctor. Then it's more about ... Okay, have I really remembered everything and asked everything that the doctor might ask about. And then I'll think that it's not my task to determine what it is, only to report what the patient says, what the patient experiences and what I see. And then I'll just insert a needle or extract a certain number of blood samples. Maybe a urine sample as well. Because I know they'll say: 'go back and get a ...[blood sample], whatever that is'.

Interviewer: So you've learned to cover your back, in order not to have to go back again?

Nurse 4: Yeah, exactly. That's usually the way it is, that that's what they'll ask for. Or if I've been with a patient and done a thorough workup, maybe I'll have some medicine that I'll leave with the patient and the patient can start treatment once I've checked with the doctor.

Nurse 1: I guess that's the way it is, you become more humble in this job the longer you do it, as you realize how little you understand. It used to be that you were very quick to think that you were right and you would deliver the workup that led to that conclusion. If you didn't have a good enough doctor, asking questions beyond what I had thought of, then the end result might be incorrect …

The experienced nurses' mindful processing of vital signs resulted in a road of action that had to be communicated and interpreted with other professionals. Another round of explicit efforts of sensemaking would take form when the nurses in this case communicated their observations and actions back to the physician.

A Shared Mind across Disciplines

Clinicians' decisions rely heavily on nurses' abilities to monitor patient symptoms and make inferences about what they hear and see and *label and report* their observations back to the physicians in ways that predispose them to find common ground. On the other hand, professionals' domain-specific knowledge must be shared with team members operating at the sharp end 'two steps away'. Specific characteristics of home care are that the 'ward' is distributed over a large geographic area and that the physicians cannot easily observe and diagnose, which raises concerns: 'How do we ensure that what I see is also what the other person sees?' This was expressed as follows:

> Of course, we always wear different lenses thanks to our differing experiences. Here, [in home care settings] I just see the patients that the nurses think I should see. The others I don't really know much about. And the way this operation is structured, it depends on that competence in order to work. There is always a risk that they don't observe things in the same way that we would if we were to see the patients – you don't get the same monitoring and quick decisions that you get if a patient is admitted to a unit.
>
> **Doctor, interview**

This was also a concern among professionals within the same discipline:

> The hard part of it is that when we give furix, [i.e. furosemide, diuretic] there are many of us doing it and it doesn't always have that much of an effect – they can remain above their goal weight for weeks and then it's important that someone reflects on it not seeming to have any effect.
>
> **Nurse, interview**

Collective sensemaking is a product of an interactive process where each professional's knowledge, experiences, values and preferences are communicated and a common ground is developed. Case 13.2 illustrates how the nurse developed a 'shared clinical eye' with the physician but was still aware of, and grounded in, her identity as a nurse. In this specific situation, being the 'extended eyes and ears' of the physician, she was able to take the physician's view and effectively bracket out essential symptoms, label them in a comprehensible way and articulate her observations back to the physician as a basis for decision-making. This is a product of sensemaking which takes time to develop within a team and also a marker of maturity in an organisation. Newly graduated nurses were aware that they had not yet achieved a 'shared mind' and compensated for that in different ways: 'I like to be on time and haven't developed the clinical eye for safety that the more experienced people have. So for me, it takes longer. I have to double check and use checklists a lot, so I am often quite late'.

Keeping Track of a Thousand Pieces

In ECW, artefacts are commonly used to share information and awareness of goals, plans and details that no single individual can grasp (Nemeth et al., 2004). An artefact could be as simple as a note on a piece of paper during a patient visit, a whiteboard in clinical areas, a worksheet, an individual nursing plan for the day-to-day work or a notebook entry during rounds. All of these are examples of artefacts used briefly, usually for a few hours, to hold and represent critical details of clinical work. Artefacts were both collective and individual, more or less tailored to the needs. As illustrated later, the selection of information for inclusion in the artefact attests to its importance and either contributes to or constrains sensemaking.

> We have the double documentation: the chart and our daily to do-list. Sometimes there's a risk that when you copy information and things are changed in the chart … who follows up on that in the to do-list? It's not that you write down exactly what to administer or what it's called or whatever … You just jot down a note on the to do-list, so you can go back and check the chart.
>
> Depending on which nurse is handing out the medications, different individual solutions are used to ensure and clarify which medications are actually handed out – if they have been replaced with a generic drug, a different strength, if they're in a pill organizer or a bag on the side. There may be post its or color coded markings on the medication lists. I use color coding, but that's just me.

Throughout the data material, it was striking that what made work hard for healthcare personnel was not encountering severely ill patients and their next of kin, handling potent drugs or technical devices. What created frustration and could make things go wrong were all the 'thousand small things' surrounding patient care. These menial tasks include ordering, storing and managing medication, providing materials to patients' homes, securing that backpacks were filled with the essentials in case of unexpected situations and documentation. This plenitude of convergent things 'to do' was captured and stored in professionals' minds during their shifts, noted down on pieces of paper or handed over in quick conversations with colleagues. Interruptions and multitasking consume memory space and cognitive ability, impeding sensemaking. Faced with complex, ambiguous and emotionally laden circumstances (cognitive overload), a person will tend to ignore data, simplify tasks and make hasty decisions and stick to rules of thumb and stereotypes rather than explore the full range of what is seen and heard. With increased complexity, the need for artefacts that facilitate sensemaking becomes pressing.

Attentiveness to 'Small Talk'

Sensemaking in home care is an ongoing process, a swift interplay of actions and interpretation in search for an answer to the question: 'what is the story?'

(Weick, 1995). Paying attention to other views and openly sharing information opens for the equally important question: 'what shall we do about it?' Sensemaking puts cognition and action together and the following case shows the ability to catch transient moments. Just being around and attentively grasping the 'small talk in the air' was of significance for the larger pattern of understanding.

Case 13.3 Attentiveness to 'Small Talk'

Mary, the coordinating nurse, makes a routine call to Mr. Carlsson to remind him to take his medication. The medical doctor, who passes by occasionally, cannot help overhearing the telephone conversation. The medical doctor, who had recently got the medical report from specialist care, is holding that paper in her hand when she passes by and the following conversation takes place between the nurse and the physician:

- 'I couldn't avoid overhearing what you said in the phone call, are we talking about the same patient?' [*showing the epicrisis*]
- 'Yes, I just talked to him and reminded him to take his medication.'
- 'But he has just been to the specialist and he should stop taking this medicine now that the treatment is complete.'
- 'Oh, so he shouldn't take it anymore … Should I call him back?'
- 'You can call the hospital (the specialist) and check if he's been given more information …'

The nurse calls the hospital and gets new information with another perspective on the patient's condition. Mary then calls Mr. Carlsson back to present the information she got from the specialist and tells him that he will get a letter sent home with more information.

This case shows how sensemaking is an issue of language, talk and communication and how engaging in sensemaking requires the ability to not only reflect and examine but also to *act* upon concrete situations.

Involving Patients in Understanding

In home care, patients and their next of kin are the hosts of care. Healthcare professionals are literally guests in a person's private home, while still having the full responsibility for medical care. Ethical dilemmas with respect to decision-making came to light in the observations. For patients living with chronic illnesses, it can be highly demanding to understand and communicate symptoms, disabilities and effects of complex regimens at home. Effectively functioning in the role of self-manager, particularly when living with one or more chronic illnesses, requires a high level of knowledge, skill and confidence. The following example illustrates how a common understanding between the nurse and the patient was achieved during a home care encounter.

Patient: What's this blue one and the white one? When you get the pills like this, you just don't know.

Nurse: Do you take those?
Patient: I have.
Nurse: Okay, but you've got them on the side.
Patient: Yeah, in a pot.
Nurse: Well, we can remove that now, because that's the Lasix. [*a medication containing furosemide, i.e. diuretic*]
Patient: Yeah, I was just wondering what the difference is.
Nurse: No, those are the same, though they can look different. This one might be green.
Patient: Yeah, I've had that one before – I just thought that was the difference, the color … That's the thing when you get medicines like this, you don't know anything. That's my collection (the nurse is looking at pills in a bag). I collect them over time and then I toss them. You have to keep track of things yourself, you know, otherwise you don't know anything …

This dialogue recognises that patients manage their health on their own most of the time, making decisions daily that affect their health. The goal of healthcare was thus to achieve not merely compliance or adherence to treatment and routines but also *sensemaking* and common grounds for decision-making.

CONCLUDING REFLECTIONS

This paper highlights six features of ECW in specialised home care, where sensemaking is a critical activity in order to cope with complexity and create safe care. These features are 'preparation, putting together the pieces of the puzzle', 'building understanding at a distance', 'a shared mind across disciplines', 'keeping track of a thousand pieces', 'attentiveness to the small talk' and 'involving patients in understanding'.

The significance of sensemaking in the *preparation* of ECW in home care is nicely illustrated in the morning preparation (Case 13.1). Sensemaking is the core activity for putting together pieces of information retrieved from team members, medical records and artefacts (e.g. notes, whiteboards) and to make 'order' out of what to an outsider could seem like 'chaos' (Weick, 1995). In order to make decisions in interaction with all these sources of information, people had to verbalise how they think about things, do things or understand things. These activities of noticing and bracketing out 'what is going on here?' with respect to a single patient's care helped the professionals reduce the plausible meanings of information and prioritise the most essential goals (Weick, 1995).

The interactions as exemplified in Case 13.1 were also imperative to exposing flaws in information transfer, to discover misconceptions between human actors as well as technical or organisational factors and to coordinate care. Perception of gaps (left corner in Figure 13.1) is central to the process of sensemaking, triggering the interplay of interpretations, inferences and actions to make the world more comprehensible. In this study, gaps are not always apparent, nor are appropriate bridging reactions always clear. The subtlest hints, like a question at a morning meeting (Case 13.1) or a spoken word that someone catches 'in the air' (Case 13.3), could be the triggers that reveal a gap present in an individual's care. This is consistent with

Weick and co-worker's (2005) view on how the order of organisations is constructed as much by the momentary and small things as by the conspicuous, large structures of, for example, written general rules.

Sensemaking goes beyond mere communication and transfer of information. It is about patterning, constructing plausible interpretations and interacting in pursuit of mutual understanding (Weick et al., 2005) (top corner of the triangle, Figure 13.1). The study shows how sensemaking rapidly switched between a *habitual mode* and a collective *sensemaking mode* when the expected and well known became unintelligible in some way. During the day, as seen in Case 13.2, sensemaking was characterised by *ongoing* noticing and bracketing of information in an inner dialogue. During this process, professionals built up an overall understanding of a patient's condition and a sensitivity to plausible roads of action. The nurses kept track of a thousand pieces of information in a clever way, using cognitive notes and artefacts, although the technical devices hindered rather than facilitated their understanding. For *expert nurses*, this process was guided by deep domain knowledge and existing mental models that had been acquired through work, training and life experience (Ekstedt and Cook, 2014), while less experienced nurses in this study described a more effortful use of cognitive skills and artefacts to make decisions on their ECW. The overall understanding of the state of patients (and the systems) which individual nurses acquired during their shift is analogous to what pilots define as 'situational awareness' or with the more practical expression of the U.S. Navy: 'having the bubble'. This means that they have sufficient expertise to see patterns and small anomalies that arise so that problems can be anticipated before they develop (Hayes, 2013). These 'bubbles' of understanding have to be shared within the team, and we found that this, much like what has been described within intensive care units (Albolino et al., 2007), was accomplished in conversations at 'rounds', handovers and morning or afternoon meetings. All personnel focused on creating a coherent, nuanced and useful understanding of the work world, its hazards, what it is possible to do and what might be the outcomes of various roads of action.

Creating collective sensemaking across disciplines is one of the most important challenges for decision-making and the development of expertise (Salas and Klein, 2001). Sensemaking is grounded in *identity construction*, which refers both to disciplinary identity and to the culture and history of each individual, which shape their thinking (Weick et al., 2005). However, the danger of blind spots in the assessment of a patient's needs in home care is increased by the fact that care, in most cases, is performed 'two steps away', making collective sensemaking even harder. There is also a delicate balance between 'shared mind' (Epstein and Street, 2011) in the sense of *becoming attenuated* and maintaining the 'professional eye' that is unique for each discipline. Both deep domain knowledge and different views are needed to perform care in complex home care settings.

In this study, the process of sensemaking was also demonstrated between the healthcare system and the patient. Several studies conclude that patients are unprepared for the self-management activities that follow hospitalisation (Moore et al., 2003; Fuji et al., 2013; Toscan et al., 2013), which requires high levels of understanding (Bodenheimer et al., 2002). A growing body of evidence shows that an active and informed patient (and family caregivers) is in most cases a basis for adherence to treatment (Stenberg et al., 2012; Flink, 2014; Wibe et al., 2014). Thus, the goal of home care organisations must be to involve patients and their next of kin in sensemaking and decision-making in the real sense of the word.

Although the organisational level was not the focus of this study, organisational learning strategies were demonstrated through the active *perception–inference–action*–interaction in *sensemaking mode*, within individual professionals and in the team. The notion of gaps is central to this process. The proposed model of sensemaking offers a link between individual and organisational learning in a dynamic (non-linear) fashion (Figure 13.1). The model suggests that management systems that offer opportunities for sensemaking and encourage people to talk about gaps in understanding foster a permissive learning atmosphere. New understanding that emerges when people interact in efforts of sensemaking during ECW becomes integrated in routines, not only on an individual professional level but also on an organisational level. This remains to be tested empirically.

ACKNOWLEDGEMENTS

The author would like to thank the Swedish Council for Health, Working life and Welfare, FORTE, for financing this project; Marlene Lindblad for her part in data collection, and Richard Cook for valuable theoretical discussions.

REFERENCES

Abraham, J., Kannampallil, T. G. and Patel, V. L. 2012. Bridging gaps in handoffs: A continuity of care based approach. *Journal of Biomedical Informatics*, 45, 240–254.

Albolino, S., Cook, R. I. and O'Connor, M. 2007. Sensemaking, safety, and cooperative work in the intensive care unit. *Cognition, Technology and Work*, 9, 131–137.

Arman, A. R., Vie, O. E. and Åsvoll, H. 2012. Refining shadowing methods for studying managerial work. In: Tengblad, S. (ed.), *The Work of Managers: Towards a Practice Theory of Management.* Oxford, U.K.: Oxford University Press, pp. 301–317.

Bodenheimer, T., Wagner, E. H. and Grumbach, K. 2002. Improving primary care for patients with chronic illness: The chronic care model, Part 2. *Journal of the American Medical Association*, 288, 1909–1914.

Caines, L. C., Brockmeyer, D. M., Tess, A. V., Kim, H., Kriegel, G. and Bates, C. K. 2011. The revolving door of resident continuity practice: Identifying gaps in transitions of care. *Journal of General Internal Medicine*, 26, 995–998.

Ciliers, P. 1998. *Complexity and Postmodernism.* London, U.K.: Routledge.

Dekker, S. 2011. *Drift into Failure. From Hunting Broken Components to Understanding Complex Systems.* Surrey, England: Ashgate.

Ekstedt, M. and Cook, R. I. 2014. The Stockholm Blizzard 2012. In: Wears, B., Hollnagel, E. and Braithwaite, J. (eds.), *The Resilience of Everyday Clinical Work.* Dorchester U.K.: Ashgate.

Epstein, R. M. and Street, R. L. 2011. Shared mind: Communication, decision making, and autonomy in serious illness. *The Annals Family Medicine*, 9, 454–461.

Fex, A., Ek, A. C. and Soderhamn, O. 2009. Self-care among persons using advanced medical technology at home. *Journal of Clinical Nursing*, 18, 2809–2817.

Flink, M. 2014. *Patients' Position in Care Transitions – An Analysis of Patient Participation and Patient-Centeredness.* Doctoral dissertation, Department of Neurobiology, Care Sciences and Society, Karolinska Institutet.

Fuji, K. T., Abbott, A. A. and Norris, J. F. 2013. Exploring care transitions from patient, caregiver, and health-care provider perspectives. *Clinical Nursing Research*, 22, 258–274.

Hayes, J. 2013. *Operational Decision-Making in High-Hazard Organizations. Drawing a Line in the Sand.* Surrey, England: Ashgate.

Jordan, M. E., Lanham, H. J., Crabtree, B. F., Nutting, P. A., Miller, W. L., Stange, K. C. and Mcdaniel, R. R. Jr. 2009. The role of conversation in health care interventions: Enabling sensemaking and learning. *Implement Science*, 4, 15.

Klein, G. J. W. 1998. *Sources of Power: How People Make Decision.* Cambridge, MA: MIT Press.

Mamykina, L., Smaldone, A. M. and Bakken, S. R. 2015. Adopting the sensemaking perspective for chronic disease self-management. *Journal of Biomedical Informatics*, 56, 406–417.

Moore, C., Wisnivesky, J., Williams, S. and Mcginn, T. 2003. Medical errors related to discontinuity of care from an inpatient to an outpatient setting. *Journal of General Internal Medicine*, 18, 646–651.

Nemeth, C., Cook, R. I., O'connor, M. and Klock, P. A. 2004. Using cognitive artifacts to understand distributed cognition. *IEEE Transactions on Systems, Man, and Cybernetics Part A: Systems and Humans*, 34, 726–735.

Nosek, J. T. 2005. Collaborative sensemaking support: Progressing from portals and tools to collaboration envelopes. *International Journal of e-Collaboration (IJeC)*, 1, 15.

Ring, P. S. and Rands, G. P. 1989. *Sensemaking, Understanding, and Committing: Emergent Interpersonal Transaction Processes in the Evolution of 3M's Microgravity Research Program*. New York: Ballinger.

Russell, D. M., Stefik, M. J., Pirolli, P. and Card, S. K. 1993. The cost structure of sensemaking. *INTERACT'93 and CHI'93 Conference on Human Factors in Computing Systems,* New York.

Salas, E. and Klein, G. (eds.), 2001. *Linking Expertise and Natural Decision Making.* Mahwah, NJ: Lawrence Erlbaum Associates Publishers.

Savolainen, R. 1993. The sense-making theory: Reviewing the interests of a user-centered approach to information seeking and use. *Information Processing & Management*, 29, 13–28.

Schutz, A. 1967. *The Phenomenology of the Social World.* Evanston, IL: Northwestern University Press.

Schutz, A. 1970. *Reflection on the Problems of Relevance.* New Haven, CT: Yale University Press.

Stenberg, U., Ruland, C. M., Olsson, M. and Ekstedt, M. 2012. To live close to a person with cancer-experiences of family caregivers. *Social Work in Health Care*, 51, 909–926.

Toscan, J., Manderson, B., Santi, S. M. and Stolee, P. 2013. Just another fish in the pond: The transitional care experience of a hip fracture patient. *International Journal of Integrated Care*, 13, e023.

Warren, B., Ballenger, C., Ogonowski, M., Rosebery, A. S. and Hudicourt-Barnes, J. 2001. Rethinking diversity in learning science. The logic of everyday sense-making. *Journal of Research Science and Technology*, 38, 529–552.

Weick, K. E. 1995. *Sensemaking in Organizations.* Thousand Oaks, CA: Sage Publications.

Weick, K. E., Sutcliffe, K. M. and Obstfeld, D. 2005. Organizing the Process of Sensemaking. *Organization Science*, 16, 409–4021.

Wibe, T., Ekstedt, M. and Helleso, R. 2015. Information practices of health care professionals related to patient discharge from hospital. *Informatics for Health & Social Care.*, 40(3), 198–209.

Woods, D. D. 1993. Process-tracing methods for the study of cognition outside of the experimental psychology laboratory. In: Klein, G., Orasanu, J., Calderwood, R. and Zsambok, C. E. (eds.), *Decision Making in Action: Models and Methods.* Norwood, NJ: Ablex, pp. 228–251.

14 Administration of Intravenous Medication

Process Variation across Hospital Wards

Eija Kivekäs, Kaisa Haatainen, Hannu Kokki and Kaija Saranto

CONTENTS

BACKGROUND

Intravenous drug administration is a more risk-prone clinical process than almost any other procedure taking place in hospital settings (Kaushal et al. 2001, Gonzales 2010). There are numerous strategies that can contribute to reducing the risk for medication errors and patient harm in the hospital drug administration processes (AAMI 2010, EU 2015). These strategies include good clinical practices, adequate training and optimal use of technologies. While literature reviews suggest that computerised infusion devices, such as smart pumps (also called smart infusion pumps or intelligent infusion devices), can contribute to the risk reduction of medication errors in healthcare settings, data supporting this manner of risk reduction in the field remain limited (Black et al. 2011, Ohashi et al. 2014).

Computerised patient infusion devices, infusion pumps, include features for preventing medication administration errors. According to earlier studies, over 90% of intravenous medications involve some type of an error (Husch et al. 2005). In Finland, a study by Ruuhilehto and her associates (2011) showed that 51% of 64,405 web-based incident reports were concerned with medication, and the most common incidents were errors in documenting, dispensing and administering medications. Valentin and colleagues (2009) found out that the administration of parenteral medication was a vulnerable area in patients' safety in intensive care. Their results were based on data from 113 participating units from all over the world (27 countries) and illustrated that this problem could not be attributed to suboptimal care in a few individual units but, instead, represented a common pattern. The most frequent errors were related to wrong time of administration and missed medication, followed by wrong dose, wrong drug and wrong route of administration. The authors stated that most medication errors occurred in routine care situation, not in extraordinary situations (Valentin et al. 2009).

Medication safety in hospitals is dependent on the successful execution of a complex system of scores of individual tasks, namely prescribing, preparing, dispensing and transcribing a medication and monitoring the patient's response. Many of these tasks lend themselves to technological devices. Patient safety is a matter of major concern. Emerging technologies, such as smart pumps, can diminish medication errors (Ohashi et al. 2013) as well as standardise and improve clinical practice, resulting in the subsequent benefits for patients (Rothschild et al. 2005, Manrique-Rodríguez et al. 2014, Mason et al. 2014).

Härkänen (2014) demonstrated that adverse outcomes due to medication-related factors were common. The use of combination of methods revealed information that was more diverse than what was previously known regarding medication-related problems in hospital setting and that can be used to increase safety in the medication process (Härkänen 2014). In incident reports and observational data, the administration errors were the most common, followed by documenting errors, while in Global Trigger Tool data, the prescribing errors were the most frequently apparent (Härkänen 2014). Technology may provide the key for reducing medical errors. Technological applications such as electronic health records (EHR), barcoding technology for medication administration and smart infusion pumps are widely held as providing solutions for patient safety (Bates and Gawande 2003, Wulff et al. 2011, Ohashi et al. 2014).

This chapter presents the findings of a study conducted in a Finnish tertiary hospital in 2014. The findings are a part of a developing process based on an international collaboration with David W. Bates and Brigham and Women's Hospital, Boston. The design section includes a description of the development project, and the results section will report on the preliminary results of a more extensive project (before–after study). The research questions focus on the following issues: what are the frequencies and types of intravenous medication errors with infusion pumps, what are the frequencies and types of intravenous medication errors within a medication process and how much variability is there by frequency and type in the settings.

SMART PUMPS: USE AND CHALLENGES

The term 'smart pump' was originally coined by the Institute for Safe Medication Practices in the United States to describe an infusion pump with an inbuilt drug library that contains correct parameters for all the medications to be delivered by the pump (ISMP 2010, Quinn 2011). Smart pump technology has progressed over the past 5 years (Ohashi et al. 2014), and the built-in safety features of the smart pump technology provide an additional double-check system for medication administration (Carayon 2010, Mason et al. 2014). The proper use of the smart pump technology yields benefits that include enhanced workflow for nurses and error reduction in medication administration (Carayon et al. 2010).

Technology provides error reduction capabilities in medication via programmed dose limit alerts with audio-visual feedback to staff regarding erroneous orders, improper dose calculations or programming errors. Ohashi and colleagues noted in their literature review lower compliance rates of using smart pumps, overriding soft alerts, non-intercepted errors and the possibility of using the wrong library. Rothschild et al. (2005) evaluated a very early version of the pumps and found that smart intravenous pumps with decision support capabilities had the capacity to intercept many dangerous medication errors. Manrique-Rodrigues et al. (2014) demonstrated that the implementation of smart pumps proved effective in preventing infusion-related programming errors from reaching patients. A study in a German hospital showed that smart pumps prevented potentially dangerous overdoses. In a before–after study on patient-controlled analgesia, the use of a smart pump resulted in a significant 22% decrease in adverse drug events recorded by an automated surveillance system, and voluntary report events also decreased significantly by 72% (Kastrup et al. 2012).

The introduction of smart pump technology provides a technology-based final check of an infusion rate that creates a safe environment for care. Modern infusion devices are much more than mere pumps and are currently often regarded as part of the organisation's information technology capability. However, only a small reduction in dosing errors has been found, while a greater reduction occurred in pump-related errors. Nine out the 10 post-intervention pump programming errors occurred because users did not use the pump software correctly (Adachi 2005, Ohashi et al. 2014). Previous studies have shown poor caregivers compliance with the drug library and users frequently ignoring the drug library when selecting a drug (Ohashi et al. 2014). Creation of a safe and effective customised drug library is essential for the proper utilisation of smart pumps. A drug library should include at least all high-alert drugs with standard concentrations as well as soft and hard stops to various dosage limits. Drug libraries must also be maintained and updated constantly. Wireless communication technology in an organisation's infrastructure allows easier adjustment or updating of drug libraries, which otherwise would require manually updating each pump separately (ISMP 2010).

REPORTING MEDICATION ERRORS' ADVERSE EVENTS: REPORTING TECHNOLOGY

The World Health Organization has presented Draft Guidelines for Adverse Event Reporting and Learning System, and a European Council Recommendation (2009/C 151/01) has also advised European countries on the establishment and revision of reporting systems (EU 2009). The European Council Recommendation mentions that there are big differences between reporting systems in the EU

Member States. There are both mandatory and voluntary reporting systems in the Member States, and variety in the types of incidents that can be reported varies. However, a broad definition allows the reporting of any concerns, including near misses and 'no harm' incidents providing a rich resource for learning and systems improvement (EU 2009).

The HaiPro system is intended for reporting patient safety incidents on the organisational level in Finland. The main properties of HaiPro are anonymity, confidentiality and freedom from sanctions. The HaiPro approach incorporates a system model that takes into consideration the features of natural human behaviour and the pathway of diverse events development. The local incident reporting system is meant to prevent adverse events of treatment through the improvement of operational procedures. At present, data are collected only at the level of the organisation and not sent forward to or nor aggregated or analysed on regional and national level (Doupi 2009). In this study, reports from the HaiPro system illustrated one perspective of the administration of intravenous medication in the hospital, and the wards were studied.

METHODS

SETTING

The study was conducted in a Finnish tertiary hospital, which has 800 beds and provides specialised medical care to 860,000 inhabitants. Around 90,000 in patients are treated in the hospital annually (PSSHP 2015). Computerised physician order entry has been in use in the hospital since 2009, and a barcoding system is used in its pharmacy. Smart pumps were in use in the intensive care unit and cardiac care unit, and a drug library was customised for each unit. The studied wards included a coronary care unit, an adult intensive care unit, an oncology unit and a cardiothoracic surgical ward. The pilot data were collected at a maternity ward, where the appropriateness of the questions was also tested for Finnish hospital.

DESIGN

This is a descriptive retrospective pre–post study conducted in three phases over the course of 36 months in total (2014–2016, Figure 14.1) based on a similar study implemented in the United States (Bates 2012). During *the first year*, the quality and quantity of reported adverse events were screened in order to acquire baseline data concerning safety incidents (including HaiPro) in the wards. This was followed by an observational study where a multidisciplinary team of investigators prospectively compared the actual medication, its dose and the infusion rate of the infusion pump with the prescription in the medical record. Subsequently, during *the second year* (2015), these results were analysed. A consensus process including a face-to-face meeting with users and decision-makers took place in order to evaluate the types of events and to develop an intervention. After a run-in period, the effectiveness of intervention will be tested during *the third year* of the study (2016) to produce a set of evidence-based recommendations.

Year 1	Year 2	Year 3
Phase 1	Phase 2	Phase 3
1. Obtain a statement of the committee on research ethics	1. Observation data analysis	1. Second measurement of intravenous medication errors
2. Development of data collection form	2. Face-to-face meeting for developing recommendations	2. Data analysis
3. Observer training	3. Interventions to reduce intravenous medication errors	3. Face-to-face meeting for developing recommendations
4. Initial measurement of intravenous medication errors		4. Publication of a final report

FIGURE 14.1 The study phases based on Bates' study protocol, 2012.

This study was approved by the University of Eastern Finland Committee on Research Ethics in 2014. Participants in the study were voluntary and aimed at patients aged 18 years or older.

Data Collection and Analysis

The observation data were collected by four researchers. The observation data collection is based on methods employed by Husch and colleagues (2005). Key elements, required to capture all kinds of intravenous medication errors, were identified using methods by Ohashi and colleagues (2013). In assigned inpatient units, using a 'prospective point prevalence' approach, all data from intravenous medications at patient bedside were collected with the standardised form. Observers compared the infusing medication dose and infusion rate on the pump with the prescribed medication, the dose and rate in the medical record. All orders were obtained from both EHR and handwritten paper-based medical records, and all intravenous fluids were classified as medication. The availability of correct patient identification band and name verification was recorded for each patient. Labelling of the infusing medication, according to medication policy, was assessed.

Observations on the intravenous medication administration processes were made over 4 days, 8 hours per day in March 2014 in each of the four units. All epidural, patient-controlled analgesia and general use of intravenous infusion pumps on inpatient care units were included in the investigation. To capture data, two to four observers (registered nurses and pharmacists) went to a ward and conducted observation on patients. The observer entered a patient room; he or she introduced himself or herself and explained the purpose of the study. In order to confirm the presence of an error (Table 14.1), the observers had to agree that an error had been made. If such an error was identified, that had the potential to cause harm, the staff nurse caring for this particular patient was discreetly informed about the incident so that he/she could correct it.

The main study outcomes will be the intravenous medication error rate and the serious medication error rate. All errors were classified for every administered medication. Serious errors were rated by NCC MERP index (National Coordinating Council for Medication Error Reporting and Prevention) (Table 14.2).

TABLE 14.1
Medication Error Types in the Study

Error Type	Definition
1. Wrong dose	The same medication but the dose is different from the prescribed order.
2. Wrong rate	A different rate is displayed on the pump from that prescribed in the medical record. Also refers to weight-based doses calculated incorrectly including using a wrong weight.
3. Wrong concentration	An amount of a medication in a unit of solution that is different from the prescribed order.
4. Wrong medication	A different fluid/medication as documented on the IV bag label is being infused compared with the order in the medical record.
5. Known allergy	Medication is prescribed/administered despite the patient had a known allergy to the drug.
6. Omitted medication	The medication ordered was not administered to a patient.
7. Delay of rate or medication/fluid change	An order to change medication or rate not carried out within 4 hours of the written order per institution policy.
8. No rate documented on label	Applies both to items sent from the pharmacy and floor stocked items per institution policy.
9. Incorrect rate on label	Rate documented on the medication label is different from that programmed into the pump. Applies to items sent from both the pharmacy and floor stocked items.
10. Patient identification error	Patient either has no ID band on wrist or information on the ID band is incorrect.
11. No documented order	Fluids/medications are being administered but no order is present in medical record. This includes failure to document a verbal order.

Source: Ohashi, K. et al., Evaluation of intravenous medication Errors with smart infusion pumps in an Academic Medical Center, AMIA, ed. in: *Proceedings of the AMIA 2013 Annual Symposium*, 16–20 November 2013, AMIA, Washington, DC, pp. 1089–1098.

TABLE 14.2
The Severity of Medication Errors according to NCC MERP Harm Index

(A) Capacity to cause error
(B) An error occurred but did not reach the patient
(C) Errors unlikely to cause harm despite reaching the patient
(D) Errors that would have required increased monitoring to preclude harm
(E) Errors likely to cause temporary harm
(F) Errors that would have caused temporary harm and prolonged hospitalisation
(G) Errors that would have produced permanent harm
(H) Errors that would have been life-threatening
(I) Errors that would likely have resulted in death

Source: NCC MERP, NCC MERP index, Homepage of National Coordinating Council for Medication Error Reporting and Prevention, 2001, Online, available at: http://www.nccmerp.org, accessed 29 January 2014.

All factors collected in each participating ward included the report of patient safety incidents at the organisational level (HaiPro), the institutional policies and procedures around medication administrations. The data were analysed and then presented at each ward. The meetings were hosted by the multidisciplinary team. Written memorandums from each meeting were analysed by content analysis, and these data will be utilised in the recommendations.

The preliminary data were used as material in a meeting with physicians, nurses, pharmacists and a patient safety manager. In the meeting, data of the use of intravenous infusion pumps and medication administration workflow were reviewed. Based on the data and the summary of the meeting, the multidisciplinary team developed an intervention that may improve the safety of intravenous drug administration at the wards.

RESULTS

The HaiPro reports from between 2011 and 2013 indicated that adverse events had increased in each unit. In particular, the incidence of near misses has increased, which illustrates that the preventative meaning of reporting has been emphasised. Most of the incidents were related to medication or information management. The majority of the reported incidents consisted of detected errors; recently, reports on near misses had increased (Figures 14.2 and 14.3).

During the data collection period, 194 inpatients in four inpatient wards were included in the study, and 492 medication procedures were observed (Table 14.3). These consisted of 355 intravenous medication infusions and 137 fluids infusions. Out of these, infusion pumps were used in 56% of cases.

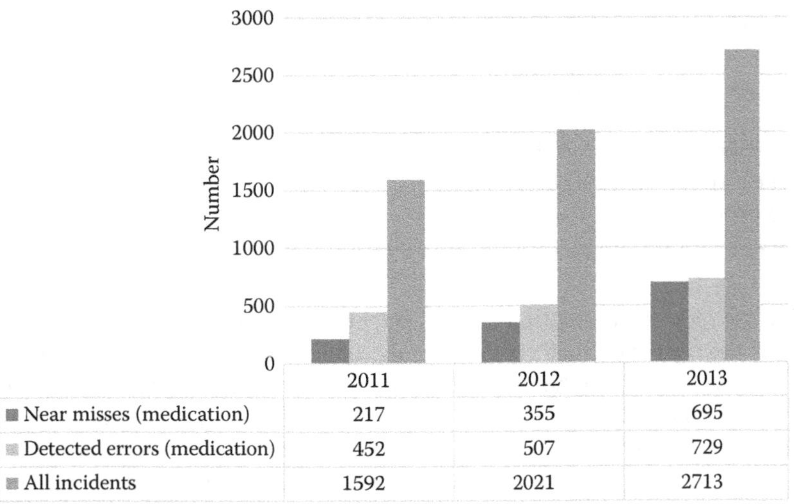

	2011	2012	2013
Near misses (medication)	217	355	695
Detected errors (medication)	452	507	729
All incidents	1592	2021	2713

FIGURE 14.2 Reported incidents of detected- and near-miss medication and intravenous medication–related errors (2011–2013) in the hospital.

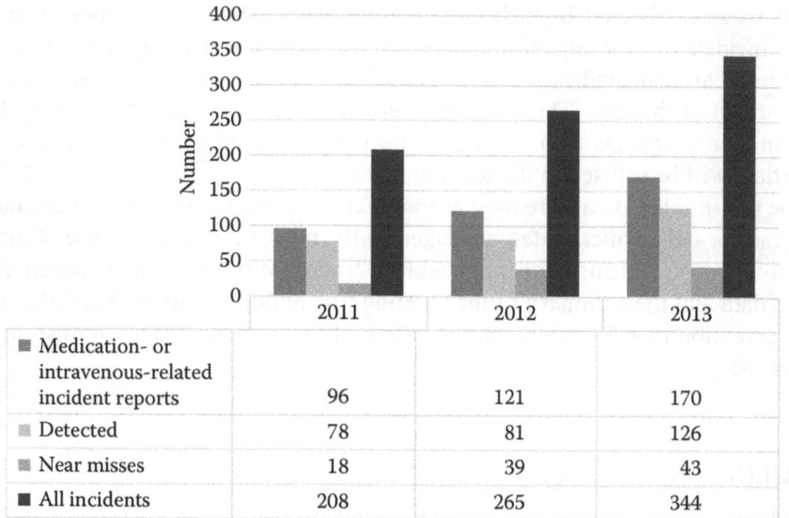

FIGURE 14.3 Reported incidents of detected- and near-miss medication and intravenous medication–related errors, 2011–2013, in the studied wards (excluding the pilot ward).

TABLE 14.3
Study Participants and Type of Medication

	Maternity Ward (a Pilot)	The Oncology Unit	The Medical ICU	The Cardiothoracic Surgical Ward	The Surgical ICU	Summary (without Pilot)
Patient	9	71	18	25	80	194
Patients who refused	1	1	2	2		5
Infusion	9	144	29	25	294	492
Medication	100%	58%	41%	80%	82%	355
Fluid		61%	59%	20%	18%	137
Infusion pump	11%	34%	83%	16%	83%	
No infusion pump	89%	60%	17%	84%	10%	
Empty medication ‘bag’		6%			20%	

The use of an identification band is imperative to ensure that the right patient receives the right care. Information on the identification band has to correspond with details recorded on the patient’s medicine chart, and this should be confirmed whenever medications are administered. Remarkable variety between wards was found in the first phase of this study (Table 14.4). Misidentification of patients and administering medicines intended for another patient to these patients are real dangers.

Stating patients’ allergies is key hospital policy. One aim of the EHR is to improve recording patients’ allergies and raise awareness of high-risk products among staff.

TABLE 14.4
Differences between Wards in the Use of Identification Bands and Documentation of Allergies

Observation Target	Wards (%/Ward)
Availability of identification band	4%–100%
Information on allergy was documented	33%–56%

TABLE 14.5
Variety between Wards in Labelling Status

Observation Target – Labelling	Wards (%/Ward)
Patient's name	2%–89%
Name of preparer (nurse or pharmacist)	0%–89%
Drug	100%
Dose or content of drug	75%–100%
Composition of infusion (volume)	8%–89%
Date/time	50%–89%

The lack of information on allergies was a surprising finding in the study (Table 14.4). Violations of medication policies regarding labelling were the most frequent error types. The details of the information missing on the labels varied between wards (Table 14.5). The most common error was the lack of information on the label of the drug to be infused. There were also the differences in the way in which medications were prescribed, and medication orders were documented. Regardless of the fact that in some wards, all prescriptions were available in the EHR, the wards used several overlapping documentation forms and a paper-based form to manage medication information.

Each error in observation data was rated using the NCC MERP harm index. Data collection in the four study units further indicated that errors were rated as A–B, indicating that there was potential for an error (A) or an error occurred but did not reach the patient (B). Indeed, in spite of the EHR system, several overlapping documentation forms were in use, and medication processes varied significantly across the units.

In summary, the work processes at each studied ward differed, even though the organisation's policy was well known. There was a lot of variety in the use of identification bands between wards, and allergies were infrequently documented. Missing allergy information was recognised as a risk in the EHR. According to the organisation policy, medication and fluid prescriptions should be documented in the EHR system. A lot of room for development was found in connection to the labels of intravenous medication, because their contents varied between wards. The use of medical technology, such as infusion pumps, smart pumps and drug library of devices, should be increased. The staff was highly interested in improving their competences in using technology.

DISCUSSION

The results of the first phase of this intervention study show the significance for the hospital to improve patient safety. There was a positive development in the voluntary reporting of patient incidents, and especially reporting near-miss incidents had increased every year. These illustrate that the preventative importance of reporting had been emphasised. Members of staff were very interested in investing in their competence in using technology and were aware of the possibilities provided by technology.

In this study, the most common error was the lack of information on medication labels thus increasing the risk of administering wrong intravenous medication to patients. Previous studies have also demonstrated similar tendencies in medication errors (Keohane et al. 2005, Ohashi et al. 2013). Documentation of allergy was the information found missing most frequently. The allergies of only 56% of the patients had been recorded in the EHR system. Differences in identifying patients in the studied wards were a further significant finding. Through the course of this study, such violation errors of hospital policy were found, which could potentially put patients at risk.

Traditional infusion pumps were in common use. Smart pumps were used mainly in cardiac care and intensive care units in this study. There was clearly a need to develop the use of technology in patient's care. However, it has already been demonstrated that until barcode pumps are integrated with other systems within the medication administration process, the role of smarts pumps in enhancing patient safety will be limited (Trbovich et al. 2010, Ohashi et al. 2014). Regardless of the technology, careful planning of care processes will ensure proper patient identification prior to any medical intervention and provide safer care with significantly fewer errors (Upton and Quinn 2013). In the future, more interest must be focused on the work processes. The work processes at each studied ward differed even though the organisation policy was well known.

The main strength of an observation and medical chart review evaluation is that the observers directly perceived the use of medical devices in clinical environments. The advantages of observation include obtaining data on authentic surroundings (Elias and Moss 2011, Sinivuo et al. 2012). The most remarkable preliminary result of this study was the high variability in the medication processes in the different studied units. Several types of errors were detected related to intravenous medication administration. These may increase the risk for administering wrong intravenous medications. The discussions in the first face-to-face meeting with the staff from the wards were productive. The staff was aware of the hospital's instructions concerning intravenous medication, but each ward had and followed their own procedures. The tendency of nurses to ignore safety software in infusion devices where manufacturers have made it optional to use these is one of the obstacles for the research efforts attempting to demonstrate and quantify an increase in patient safety (Rothschild et al. 2005). The temptation to follow so-called workarounds must be resisted if potential benefits to patients and health professionals are to be realised (McAlearney et al. 2007).

In the meetings with wards staff, the significance of the problem was recognised, and deeper commitment to hospital policy was stated. Overlapping documentation and ward-specific systems were highlighted, and there were lively discussions on the recommendations. Collaboration in the hospital was emphasised. It is the prime responsibility of the prescriber to ensure that the patient's information is checked for allergies before any prescriptions are written. Other staff responsible for dispensing, administrating and monitoring medicines can help prevent potential incidents by identifying the patient's allergy status.

The results of discussion with the staff have condensed into four recommendations: (1) the identification band should be used according to the organisation policy; (2) allergy status should be documented for all patients, including the lack of an allergy ('No known medicine allergy'); (3) medication and fluid orders should be prescribed on the EHR system according to the organisation policy and (4) the label of intravenous medication has to be filled as instructed.

There are limitations to our study. The small sample size of the studied wards limits the generalisability of the results. The study focused only on one tertiary hospital. However, as they are in use all over healthcare institutions, the safe use of infusion pumps is a shared challenge for healthcare organisations.

The first observations on intravenous medication errors identified the key issues related to the use of infusion pumps and smart pumps. This will support to the development of strategies to allow improving the prevention of intravenous medication errors. The utilisation of staff meetings will increase commitment to develop strategies for improving the prevention of intravenous medication errors. Support gained from hospital management and their commitment to improve patient safety was a facilitator for this study.

CONCLUSION

The most important preliminary result of this first observation study was the high variability in the medication processes in the studied wards. The hospital policy concerning intravenous medication has a central role in improving patient safety. The staff appeared to be well motivated to develop the medication processes and their own competence in using technology. Future challenges associated with the hospital policy especially to intravenous medication and increasing the use of medical technologies in patient care and in healthcare in general.

REFERENCES

AAMI, 2010-last update, Infusing patients safely [Homepage of The Association for the Advancement of Medical Instrumentation (AAMI)], [Online]. Available at: http://s3.amazonaws.com/rdcms-aami/files/production/public/FileDownloads/Summits/AAMI_FDA_Summit_Report.pdf (accessed 20 May 2015).

Adachi, W. and Lodolce, A.E., 2005. Use of failure mode and effects analysis in improving the safety of i.v. drug administration. *American Journal of Health-System Pharmacy*, 62, 917–920.

Bates, D.W., 2012. Research proposal. A national study of intravenous medication errors: Understanding how to improve intravenous safety with smart pumps. Brigham and Women's Hospital, Boston, MA (unprinted).

Bates, D.W. and Gawande, A.A., 2003. Improving safety with information technology. *The New England Journal of Medicine*, 248, 2526–2534.

Black, A.D., Car, J., Pagliari, C., Cresswell, K., Bokum, T., Mckinstry, B., Procter, R., Majeed, A. and Sheikh, A., 2011. The impact of eHealth on the quality and safety of health care: A systematic overview. *PLoS Medicine*, 8, e1000387.

Carayon, P., Hundt, A.S. and Wetterneck, T.B., 2010. Nurses' acceptance of Smart IV pump technology. *International Journal of Medical Informatics*, 79(6), 401–411.

Doupi, P., 2009. *National Reporting Systems for Patient Safety Incidents. A Review of the Situation in Europe*. Jyväskylä, Finland: National Institute for Health and Welfare.

Elias, B.L. and Moss, J.A., 2011. Smart pump technology. What we have learned. *Computers, Informatics, Nursing*, 29, 184–190.

EU, 2009, Council Recommendation of 9 June 2009 on patient safety, including the prevention and control of healthcare associated infections (2009/C 151/01) [Homepage of European Commission], [Online]. Available at: http://ec.europa.eu/health/patient_safety/docs/council_2009_en.pdf (accessed 12 August 2014).

EU, 2015–last update, Report on the public consultation on patient safety and quality of care [Homepage of European Commission, Patient Safety and Quality of Care working group], [Online]. Available at: http://ec.europa.eu/health/patient_safety/docs/pasq_public_consultation_report.pdf (accessed 12 August 2015).

Gonzales, K., 2010. Medication administration errors and the pediatric population: A systematic search of the literature. *Journal of Pediatric Nursing*, 25, 555–565.

Härkänen, M., 2014. Medication-related adverse outcomes and contributing factors among hospital patients an analysis using hospital's incident reports, the global trigger tool method, and observations with record reviews, The Faculty of Health Sciences, University of Eastern Finland, Finland.

Husch, M., Sullivan, C., Rooney, D., Barnard, C., Fotis, M., Clarke, J. and Noskin, G., 2005. Insights from the sharp end of intravenous medication errors: Implications for infusion pump technology. *Quality and Safety in Health Care*, 14, 80–86.

ISMP, 2010. *Multiple Intravenous Infusions Phase 1a: Situation Scan Summary Report*. Toronto, Canada: ISMP.

Kastrup, M., Balzer, F., Volk, T. and Spies, C., 2012. Analysis of event logs from syringe pumps a retrospective pilot study to assess possible effects of syringe pumps on safety in a University Hospital Critical Care Unit in Germany. *Drug Safety*, 35(7), 563–574.

Kaushal, R., Bates, D.W., Landrigan, C., Mckenna, K.J., Clapp, M.D., Federico, F. and Goldmann, D.A., 2001. Medication errors and adverse drug events in pediatric inpatients. The *Journal of the American Medical Association*, 285, 2114–2120.

Keohane, C.A., Hayes, J., Saniuk, C., Rothschild, J.M. and Bates, D.W., 2005. Intravenous medication safety and smart infusion systems. Lesson learned and future opportunities. *Journal of Infusion Nursing*, 28, 321–328.

Manrique-Rodríguez, S., Sánchez-Galindo, A.C., López-Herce, J., Calleja-Hernández, M.Á, Martínez-Martínez, F., Iglesias-Peinado, I., Carrillo-Álvarez, Á, Sanjurjo-Sáez, M. and Fernández-Llamazares, C.M., 2014. Implementing smart pump technology in a pediatric intensive care unit: A cost-effective approach. *International Journal of Medical Informatics*, 83(2), 99–105.

Mason, J.J., Roberts-Turner, R. and Hinds, P.S., 2014. Patient safety, error reduction, and pediatric nurses' perceptions of smart pump technology. *Journal of Pediatric Nursing*, 29, 143–151.

McAlearney, A.S., Vrontos, J., Schneider, P.J., Curran, C.R., Czerwinski, B.S., Pedersen, C.A., 2007. Strategic work-arounds to accommodate new technology: The case of smart pumps in hospital care. *Journal of Patient Safety*, 3, 75–81.

NCC MERP, 2001. NCC MERP index [Homepage of National Coordinating Council for Medication Error Reporting and Prevention], [Online]. Available at: http://www.nccmerp.org (accessed 29 January 2014).

Ohashi, K., Dalleur, O., Dykes, P.C. and Bates, D.W., 2014. Benefits and risks of using smart pumps to reduce medication error rates: A systematic review. *Drug Safety*, 37, 1011–1020.

Ohashi, K., Dykes, P., Mcintosh, K., Buckley, E., Wien, M. and Bates D.W., 2013. Evaluation of intravenous medication Errors with smart infusion pumps in an Academic Medical Center, AMIA, Ed. In: *Proceedings of the AMIA 2013 Annual Symposium*, 16–20 November 2013, Washington, DC: AMIA, pp. 1089–1098.

PSSHP, 2015 last update, Kuopio University Hospital [Homepage of University Hospital], [Online]. Available at: https://www.psshp.fi/web/en/ (accessed 12 August 2015).

Quinn, C., 2011. Smart practice: The introduction of a dose error reduction system. *British Journal of Nursing*, 20, S20–S25.

Rothschild, J.M., Keohane, C.A., Cook, F.E., Orav, J., Burdick, E., Thompson, S., Hayes, J. and Bates, D.W., 2005. A controlled trial of smart infusion pumps to improve medication safety in critically ill patients. *Critical Care Medicine*, 33, 533–540.

Sinivuo, R., Koivula, M. and Kylmä, J., 2012. Observation as a data collection methods in clinical context (in Finnish). *Journal of Nursing Science*, 24, 291–301.

Trbovich, P.L., Pinkney, S., Cafazzo, J.A., Easty, A.C., 2010. The impact of traditional and smart pump infusion technology on nurse medication administration performance in a simulated inpatient unit. *Quality and Safety in Health Care*, 19, 430–434.

Upton, D. and Quinn, C., 2013. Smart pumps – Good for nurses as well as patients. *British Journal of Nursing*, 22, 4–8.

Valentin, A., Capuzzo, M., Guidet, B., Moreno, R., Metnitz, B., Bauer, P. and Metnitz, P., 2009. Errors in administration of parenteral drugs in intensive care units: Multinational prospective study. *British Medical Journal*, 338, b814.

Wulff, K., Cummings, G., Marck, P. and Yurtseven, O., 2011. Medication administration technologies and patient safety: A mixed-method systematic review. *Journal of Advanced Nursing*, 67, 2080–2095.

Appendix

STUDY ID

Aase, K. et al., Patient safety challenges in a case study hospital – Of relevance for transfusion processes? *Transfusion and Apheresis Science*, 2008. 39(2): 167–172.

Aasen, E.M., A comparison of the discursive practices of perception of patient participation in haemodialysis units. *Nursing Ethics*, 2014.

Adler-Milstein, J. et al., Benchmarking health IT among OECD countries: Better data for better policy. *Journal of the American Medical Informatics Association*, 2014. 21(1): 111–116.

Adolfsson, E.T., A. Rosenblad and K. Wikblad, The Swedish national survey of the quality and organization of diabetes care in primary healthcare – Swed-QOP. *Primary Care Diabetes*, 2010. 4(2): 91–97.

Agvall, B., U. Alehagen and U. Dahlström, The benefits of using a heart failure management programme in Swedish primary healthcare. *European Journal of Heart Failure*, 2013. 15(2): 228–236.

Ahlén, G.C. and R.K. Gunnarsson, The physician's self-evaluation of the consultation and patient outcome: A longitudinal study. *Scandinavian Journal of Primary Health Care*, 2013. 31(1): 26–30.

Alenius, L.S. et al., Staffing and resource adequacy strongly related to Rns' assessment of patient safety: A national study of Rns working in acute-care hospitals in Sweden. *BMJ Quality and Safety*, 2014. 23(3): 242–249.

Ammendrup, A.C. et al., Urinary incontinence surgery in Denmark 2001–2003. *Ugeskrift For Laeger*, 2009a. 171(6): 399–404.

Ammendrup, A.C. et al., A Danish national survey of women operated with mid-urethral slings in 2001. *Acta Obstetricia et Gynecologica Scandinavica*, 2009b. 88(11): 1227–1233.

Ammentorp, J. et al., Measurement of the quality of care in a paediatric department. *Ugeskrift for Laeger*, 2001. 163(50): 7048–7052.

Andersson, A.C. et al., Evaluating a questionnaire to measure improvement initiatives in Swedish healthcare. *BMC Health Services Research*, 2013. 13(1).

André, B. et al., The impact of work culture on quality of care in nursing homes – A review study. *Scandinavian Journal of Caring Sciences*, 2013.

Arpin, S., Oral hygiene in elderly people in hospitals and nursing homes. *Evidence-Based Dentistry*, 2009. 10(2): 46.

Bakken, K. et al., Tidsskrift For Den Norske Lægeforening: Tidsskrift For Praktisk Medicin, Ny Række, [Insufficient communication and information regarding patient medication in the primary healthcare]. 2007. 127(13): 1766–1769.

Berg, L.M. et al., Interruptions in emergency department work: An observational and interview study. *BMJ Quality and Safety*, 2013. 22(8): 656–663.

Berland, A., D. Gundersen and S.B. Bentsen, Patient safety and falls: A qualitative study of home care nurses in Norway. *Nursing and Health Sciences*, 2012. 14(4): 452–457.

Bernstein, K., M. Bruun-Rasmussen and S. Vingtoft. A method for specification of structured clinical content in electronic health records. *Studies in Health Technology and Informatics*, 2006.

Björkenstam, E. et al., Quality of medical care and excess mortality in psychiatric patients – A nationwide register-based study in Sweden. *BMJ Open*, 2012. 2: e000778–e000778.

Bjorkman, M. and K. Malterud, Lesbian women's experiences with health care: A qualitative study. *Scandinavian Journal of Primary Health Care*, 2009. 27(4): 238–243.

Bjørn, B., J. Anhøj and B. Lilja, Reporting of patient safety incidents: Experience from five years with a national reporting system. *Ugeskrift For Laeger*, 2009. 171(20): 1677–1680.

Bjørnstad, C.C.L. et al., Temporary cardiac pacemaker treatment in five Norwegian regional hospitals. *Scandinavian Cardiovascular Journal*, 2012. 46(3): 137–143.

Bolse, K., I. Thylén and A. Strömberg, Healthcare professionals' experiences of delivering care to patients with an implantable cardioverter defibrillator. *European Journal of Cardiovascular Nursing*, 2013. 12(4): 346–352.

Bostrom, A.M. et al., Factors associated with evidence-based practice among registered nurses in Sweden: A national cross-sectional study. *BMC Health Services Research*, 2013. 13: 165.

Brännström, M. et al., Unequal care for dying patients in Sweden: A comparative registry study of deaths from heart disease and cancer. *European Journal of Cardiovascular Nursing: Journal of the Working Group on Cardiovascular Nursing of the European Society of Cardiology*, 2012. 11(4): 454–459.

Brattheim, B., A. Faxvaag and A. Seim, Process support for risk mitigation: A case study of variability and resilience in vascular surgery. *BMJ Quality & Safety*, 2011. 20(8): 672–679.

Browall, M. et al., Patients' experience of important factors in the healthcare environment in oncology care. *International Journal of Qualitative Studies on Health & Well-Being*, 2013. 8: 1–10.

Brüggemann, A.J., B. Wijma and K. Swahnberg, Patients' silence following healthcare staff's ethical transgressions. *Nursing Ethics*, 2012. 19(6): 750–763.

Bylund, C., Structured patient-clinician communication using DIALOG improves patient quality of life. *Evidence Based Mental Health*, 2008. 11(3): 89.

Bystedt, M., M. Eriksson and B. Wilde-Larsson, Delegation within municipal health care. *Journal of Nursing Management*, 2011. 19(4): 534–541.

Christiaens, T. et al., Guidelines, evidence, and cultural factors. *Scandinavian Journal of Primary Health Care*, 2004. 22(3): 141–145.

de Beaufort, C. et al., Harmonize care to optimize outcome in children and adolescents with diabetes mellitus: Treatment recommendations in Europe. *Pediatric Diabetes*, 2012. 13(Suppl 16): 15–19.

Drösler, S.E. et al., Application of patient safety indicators internationally: A pilot study among seven countries. *International Journal for Quality in Health Care: Journal of the International Society for Quality in Health Care/Isqua*, 2009. 21(4): 272–278.

Eriksson, H. et al., Reducing queues: Demand and capacity variations. *International Journal of Health Care Quality Assurance*, 2011. 24(8): 592–600.

Forsberg, E., R. Axelsson and B. Arnetz, Effects of performance â-based reimbursement in healthcare. *Scandinavian Journal of Public Health*, 2000. 28(2): 102–110.

Fossum, M. et al., Translation and testing of the Risk Assessment Pressure Ulcer Sore scale used among residents in Norwegian nursing homes. *BMJ Open*, 2012. 2(5).

Fossum, M. et al., Effects of a computerized decision support system on care planning for pressure ulcers and malnutrition in nursing homes: An intervention study. *International Journal of Medical Informatics*, 2013.

Frank, C. et al., Questionnaire for patient participation in emergency departments: Development and psychometric testing. *Journal of Advanced Nursing*, 2011a. 67(3): 643–651.

Frank, C. et al., Patient participation in the emergency department: An evaluation using a specific instrument to measure patient participation (PPED). *Journal of Advanced Nursing*, 2011b. 67(4): 728–735.

Fröjd, C. et al., Patient information and participation still in need of improvement: Evaluation of patients' perceptions of quality of care. *Journal of Nursing Management*, 2011. 19(2): 226–236.

Frølich, A. et al., Integration of healthcare rehabilitation in chronic conditions. *International Journal of Integrated Care*, 2010. 10: e033.

Garson, A. et al., International differences in patient and physician perceptions of "High Quality" healthcare: A model from pediatric cardiology. *American Journal of Cardiology*, 2006. 97(7): 1073–1075.

Grøndahl, V.A. et al., Quality of care from patients' perspective: Impact of the combination of person-related and external objective care conditions. *Journal of Clinical Nursing*, 2011. 20(17–18): 2540–2551.

Gunningberg, L., M. Fogelberg-Dahm and A. Ehrenberg, Improved quality and comprehensiveness in nursing documentation of pressure ulcers after implementing an electronic health record in hospital care. *Journal of Clinical Nursing*, 2009. 18(11): 1557–1564.

Häggström, M., K. Asplund and L. Kristiansen, Important quality aspects in the transfer process. *International Journal of Health Care Quality Assurance*, 2014. 27(2): 123–139.

Hajdu, A. et al., Evaluation of the national surveillance system for point-prevalence of healthcare-associated infections in hospitals and in long-term care facilities for elderly in Norway, 2002–2008. *BMC Public Health*, 2011. 11: 923.

Hakkarainen, K.M. et al., Prevalence and perceived preventability of self-reported adverse drug events – A population-based survey of 7099 adults. *PLoS ONE*, 2013. 8(9).

Hakkarainen, K.M. et al., Prevalence, nature and potential preventability of adverse drug events – A population-based medical record study of 4970 adults. *British Journal of Clinical Pharmacology*, 2014. 78(1): 170–183.

Håkonsen, H. et al., Generic substitution: A potential risk factor for medication errors in hospitals. *Advances in Therapy*, 2010. 27(2): 118–126.

Härenstam, K.P. et al., Patient safety as perceived by Swedish leaders. *International Journal of Health Care Quality Assurance*, 2009. 22(2): 168–182.

Härkänen, M. et al., Medication errors: What hospital reports reveal about staff views. *Nursing Management*, 2013. 19(10): 32–37.

Hätönen, H. et al., Patient education practices in psychiatric hospital wards: A national survey in Finland. *Nordic Journal of Psychiatry*, 2010. 64(5): 334–339.

Hedsköld, M. et al., Psychometric properties of the Hospital Survey on Patient Safety Culture, HSOPSC, applied on a large Swedish health care sample. *BMC Health Services Research*, 2013. 13: 332.

Helmiö, P. et al., WHO Surgical Safety Checklist in otorhinolaryngology-head and neck surgery: Specialty-related aspects of check items. *Acta Oto-Laryngologica*, 2012. 132(12): 1334–1341.

Hessén Söderman, A.C. et al., Reduced risk of primary postoperative hemorrhage after tonsil surgery in Sweden: Results from the national tonsil surgery register in Sweden covering more than 10 years and 54,696 operations. *Laryngoscope*, 2011. 121(11): 2322–2326.

Hofoss, D. and E. Deilkås, Roadmap for patient safety research: Approaches and roadforks. *Scandinavian Journal of Public Health*, 2008. 36(8): 812–817.

Hollman, D., S. Lennartsson and K. Rosengren, District nurses' experiences with the free-choice system in Swedish primary care. *British Journal of Community Nursing*, 2014. 19(1): 30–35.

Hove, L.D., J. Bock and J.K. Christoffersen, Analysis of deaths among children in the period 1996–2008 from closed claims registered by the Danish Patient Insurance Association. *Acta Paediatrica, International Journal of Paediatrics*, 2012. 101(10): 1074–1078.

Hovlid, E. and O. Bukve, A qualitative study of contextual factors' impact on measures to reduce surgery cancellations. *BMC Health Services Research*, 2014. 14: 215.

Hovlid, E. et al., Sustainability of healthcare improvement: What can we learn from learning theory? *BMC Health Services Research*, 2012. 12(1).

Ingemansson, M. et al., Adherence to guidelines for drug treatment of asthma in children: Potential for improvement in Swedish primary care. *Quality in Primary Care*, 2012. 20(2): 131–139.

Isola, A. et al., Quality of institutional care of older people as evaluated by nursing staff. *Journal of Clinical Nursing*, 2008. 17(18): 2480–2489.

Itoh, K. et al., Patient views of adverse events: Comparisons of self-reported healthcare staff attitudes with disclosure of accident information. *Applied Ergonomics*, 2006. 37(4 SPEC. ISS.): 513–523.

Iversen, H.H., O.A. Bjertnaes and K.E. Skudal, Patient evaluation of hospital outcomes: An analysis of open-ended comments from extreme clusters in a national survey. *BMJ Open*, 2014. 4(5): e004848.

Jakobsen, D.H. et al., Standardising fast-track surgical nursing care in Denmark. *British Journal of Nursing*, 2014. 23(9): 471–476.

Jakobsson, J. and C. Wann-Hansson, Nurses' perceptions of working according to standardized care plans: A questionnaire study. *Scandinavian Journal of Caring Sciences*, 2013. 27(4): 945–952.

Jakobsson, L. and L. Holmberg, Quality from the patient's perspective: A one-year trial. *International Journal of Health Care Quality Assurance*, 2012. 25(3): 177–188.

Jensen, N.K., S.S. Nielsen and A. Krasnik, Expert opinion on "best practices" in the delivery of health care services to immigrants in Denmark. *Danish Medical Bulletin*, 2010. 57(8): A4170.

Johansen, B. et al., Quality improvement in an outpatient department for subacute low back pain patients: Prospective surveillance by outcome and performance measures in a health technology assessment perspective. *Spine*, 2004. 29(8): 925–931.

Johansen, I. and M. Rasmussen, Electronic requests for laboratory tests by general practitioners greatly reduce errors and costs. *Journal on Information Technology in Healthcare*, 2009. 7(1): 49–57.

Johansson, A., B. Andershed and A. Anderzen-Carlsson, Conceptions of mental health care – From the perspective of parents' of adult children suffering from mental illness. *Scandinavian Journal of Caring Sciences*, 2013.

Johansson, M.-L., S. Hägg and S.M. Wallerstedt, Impact of information letters on the reporting rate of adverse drug reactions and the quality of the reports: A randomized controlled study. *BMC Clinical Pharmacology*, 2011. 11: 14.

Johnson, J.K., V.M. Arora and P.R. Barach, What can artefact analysis tell us about patient transitions between the hospital and primary care? Lessons from the HANDOVER project. *European Journal of General Practice*, 2013. 19(3): 185–193.

Kaakinen, P., H. Kyngäs and M. Kääriäinen, Predictors of good-quality counselling from the perspective of hospitalised chronically ill adults. *Journal of Clinical Nursing*, 2013. 22(19/20): 2704–2713.

Källberg, A.S. et al., Medical errors and complaints in emergency department care in Sweden as reported by care providers, healthcare staff, and patients – A national review. *European Journal of Emergency Medicine*, 2013. 20(1): 33–38.

Källman, U. and B.O. Suserud, Knowledge, attitudes and practice among nursing staff concerning pressure ulcer prevention and treatment – A survey in a Swedish healthcare setting. *Scandinavian Journal of Caring Sciences*, 2009. 23(2): 334–341.

Khorram-Manesh, A., A. Hedelin and P. Ortenwall, Hospital-related incidents; causes and its impact on disaster preparedness and prehospital organisations. *Scandinavian Journal of Trauma, Resuscitation and Emergency Medicine*, 2009. 17: 26.

Kihlgren, A.L., M. Nilsson and V. Sørlie, Caring for older patients at an emergency department – Emergency nurses' reasoning. *Journal of Clinical Nursing*, 2005. 14(5): 601–608.

Kivekäs, E. et al., Improving the coordination of patients' medication management: A regional Finnish development project. *Studies in Health Technology and Informatics*, 2014. 201: 175–180.

Kjaer, M.L. et al., Venous leg ulcer patient priorities and quality of care: Results of a survey. *Ostomy/Wound Management*, 2004. 50(1): 48–55.

Kousgaard, M.B., A.S. Joensen and T. Thorsen, Reasons for not reporting patient safety incidents in general practice: A qualitative study. *Scandinavian Journal of Primary Health Care*, 2012. 30(4): 199–205.

Kristoffersen, A.H., G. Thue and S. Sandberg, Postanalytical external quality assessment of warfarin monitoring in primary healthcare. *Clinical Chemistry*, 2006. 52(10): 1871–1878.

Kuusela, M. et al., The medico-professional quality of GP consultations assessed by analysing patient records. *Scandinavian Journal of Primary Health Care*, 2011. 29(4): 222–226.

Kværner, K.J., Benchmarking surgery: Secondary post-tonsillectomy hemorrhage 1999–2005. *Acta Oto-Laryngologica*, 2009. 129(2): 195–198.

Kværner, K.J. et al., Hospital discharge information as a communication tool. *Tidsskrift for den Norske Laegeforening*, 2005. 125(20): 2815–2817.

Lipczak, H., J.L. Knudsen and A. Nissen, Safety hazards in cancer care: Findings using three different methods. *BMJ Quality and Safety*, 2011. 20(12): 1052–1056.

Magrabi, F. et al., A comparative review of patient safety initiatives for national health information technology. *International Journal of Medical Informatics*, 2013. 82(5): e139–e148.

Martin, H.M., L.E. Navne and H. Lipczak, Involvement of patients with cancer in patient safety: A qualitative study of current practices, potentials and barriers. *BMJ Quality & Safety*, 2013. 22(10): 836–842.

Mattsson, T.O. et al., Assessment of the global trigger tool to measure, monitor and evaluate patient safety in cancer patients: Reliability concerns are raised. *BMJ Quality and Safety*, 2013. 22(7): 571–579.

Nørgaard, B. et al., Communication skills training for health care professionals improves the adult orthopaedic patient's experience of quality of care. *Scandinavian Journal of Caring Sciences*, 2012. 26(4): 698–704.

Öhrn, A. and G. Eriksson. Risk analysis – A tool for IT development and patient safety a comparative study of weaknesses before and after implementation of a health care system in the county council of ostergotland, Sweden. *Studies in Health Technology and Informatics*, 2007.

Oja, P., T. Kouri and A. Pakarinen, Health centres' view of the services provided by a university hospital laboratory: Use of satisfaction surveys. *Scandinavian Journal of Primary Health Care*, 2010. 28(1): 24–28.

Oksuzyan, A. et al., Changes in hospitalisation and surgical procedures among the oldest-old: A follow-up study of the entire Danish 1895 and 1905 cohorts from ages 85 to 99 years. *Age & Ageing*, 2013. 42(4): 476–481.

Olsen, E. and K. Aase, A comparative study of safety climate differences in healthcare and the petroleum industry. *Quality & Safety in Health Care*, 2010. 19(Suppl 3): i75–i79.

Olsson, I.N., R. Runnamo and P. Engfeldt, Drug treatment in the elderly: An intervention in primary care to enhance prescription quality and quality of life. *Scandinavian Journal of Primary Health Care*, 2012. 30(1): 3–9.

Oltedal, S. et al., The NORPEQ patient experiences questionnaire: Data quality, internal consistency and validity following a Norwegian inpatient survey. *Scandinavian Journal of Public Health*, 2007. 35(5): 540–547.

Øvretveit, J., Quality evaluation and indicator comparison in health care. *International Journal of Health Planning and Management*, 2001. 16(3): 229–241.

Øvretveit, J., Producing useful research about quality improvement. *International Journal of Health Care Quality Assurance*, 2002. 15(7): 294–302.

Øvretveit, J. and M.A. Sachs, What is the cost of the patient safety and quality problem in healthcare? A review of the research. *Lakartidningen*, 2005. 102(3): 140–142.

Øvretveit, J. et al., Improving quality through effective implementation of information technology in healthcare. *International Journal for Quality in Health Care*, 2007. 19(5): 259–266.

Palmqvist, E. et al., Difficulties for primary health care staff in interpreting bacterial findings on a device for simplified urinary culture. *Scandinavian Journal of Clinical and Laboratory Investigation*, 2008. 68(4): 312–316.

Papastavrou, E. et al., A cross-cultural study of the concept of caring through behaviours: Patients' and nurses' perspectives in six different EU countries. *Journal of Advanced Nursing*, 2012. 68(5): 1026–1037.

Pedersen, R. et al., In quest of justice? Clinical prioritisation in healthcare for the aged. *Journal of Medical Ethics*, 2008. 34(4): 230–235.

Petersson, I.F. et al., Development of healthcare quality indicators for rheumatoid arthritis in Europe: The eumusc.net project. *Annals of the Rheumatic Diseases*, 2014. 73(5): 906–908.

Putzer, G. et al., LUCAS compared to manual cardiopulmonary resuscitation is more effective during helicopter rescue-a prospective, randomized, cross-over manikin study. *The American Journal of Emergency Medicine*, 2013. 31(2): 384–389.

Rabøl, L.I. et al., Descriptions of verbal communication errors between staff. An analysis of 84 root cause analysis-reports from Danish hospitals. *BMJ Quality and Safety*, 2011a. 20(3): 268–274.

Rabøl, L.I. et al., Republished error management: Descriptions of verbal communication errors between staff. An analysis of 84 root cause analysis-reports from Danish hospitals. *Postgraduate Medical Journal*, 2011b. 87(1033): 783–789.

Rahimi, B. et al., Implementing an integrated computerized patient record system: Towards an evidence-based information system implementation practice in healthcare. AMIA… Annual Symposium Proceedings. *AMIA Symposium*, 2008: 616–620.

Rahimi, B. et al., Organization-wide adoption of computerized provider order entry systems: A study based on diffusion of innovations theory. *BMC Medical Informatics and Decision Making*, 2009. 9(1).

Ranji, A., A.-K. Dykes and P. Ny, Routine ultrasound investigations in the second trimester of pregnancy: The experiences of immigrant parents in Sweden. *Journal of Reproductive & Infant Psychology*, 2012. 30(3): 312–325.

Rask, M.T. et al., Effects of an intervention aimed at improving nurse-patient communication in an oncology outpatient clinic. *Cancer Nursing*, 2009. 32(1): E1–E11.

Robertsson, O., J. Ranstam and L. Lidgren, Variation in outcome and ranking of hospitals: An analysis from the Swedish knee arthroplasty register. *Acta Orthopaedica*, 2006. 77(3): 487–493.

Röing, M., U. Rosenqvist and I. K. Holmström, Threats to patient safety in telenursing as revealed in Swedish telenurses' reflections on their dialogues. *Scandinavian Journal of Caring Sciences*, 2013. 27(4): 969–976.

Rokstad, I.S. et al., Electronic optional guidelines as a tool to improve the process of referring patients to specialized care: An intervention study. *Scandinavian Journal of Primary Health Care*, 2013. 31(3): 166–171.

Rytter, L. et al., Comprehensive discharge follow-up in patient's homes by GPs and district nurses of elderly patients. *Scandinavian Journal of Primary Health Care*, 2010. 28(3): 146–153.

Sampo, M.M. et al., Soft tissue sarcoma – A population-based, nationwide study with special emphasis on local control. *Acta Oncologica*, 2012. 51(6): 706–712.

Schildmeijer, K. et al., Assessment of adverse events in medical care: Lack of consistency between experienced teams using the global trigger tool. *BMJ Quality & Safety*, 2012. 21(4): 307–314.

Schildmeijer, K. et al., Retrospective record review in proactive patient safety work – Identification of no-harm incidents. *BMC Health Services Research*, 2013. 13(1).

Schiøtz, M. et al., Something is amiss in Denmark: A comparison of preventable hospitalisations and readmissions for chronic medical conditions in the Danish Healthcare system and Kaiser Permanente. *BMC Health Services Research*, 2011. 11: 347.

Schmidt, S. et al., Healthcare needs and healthcare satisfaction from the perspective of parents of children with chronic conditions: The DISABKIDS approach towards instrument development. *Child: Care, Health and Development*, 2008. 34(3): 355–366.

Siemsen, I.M.D. et al., Factors that impact on the safety of patient handovers: An interview study. *Scandinavian Journal of Public Health*, 2012. 40(5): 439–448.

Siersma, V., G. Waldemar and F.B. Waldorff, Subjective memory complaints in primary care patients and death from all causes: A four-year follow-up. *Scandinavian Journal of Primary Health Care*, 2013. 31(1): 7–12.

Sinnemäki, J. et al., Automated dose dispensing service for primary healthcare patients: A systematic review. *Systematic Reviews*, 2013. 2: 1.

Sipilä, R. et al., Facilitating as a guidelines implementation tool to target resources for high risk patients – The Helsinki Prevention Programme (HPP). *Journal of Interprofessional Care*, 2008. 22(1): 31–44.

Sjøhart Lund, M.L. et al., Quality of outpatient tonsillectomy performed in ear, nose & throat practice. *Ugeskrift for Laeger*, 2010. 172(28).

Skålén, P. et al., The contextualization of human resource and quality management: A sensemaking perspective on everybody's involvement. *International Journal of Human Resource Management*, 2005. 16(5): 736–751.

Skär, L. and S. Söderberg, Complaints with encounters in healthcare – Men's experiences. *Scandinavian Journal of Caring Sciences*, 2012. 26(2): 279–286.

Skarstein, J. et al., 'Patient satisfaction' in hospitalized cancer patients. *Acta Oncologica*, 2002. 41(7/8): 639–645.

Skudal, K.E. et al., The Nordic Patient Experiences Questionnaire (NORPEQ): Cross-national comparison of data quality, internal consistency and validity in four Nordic countries. *BMJ Open*, 2012. 2(3).

Smith, F. et al., Readability, suitability and comprehensibility in patient education materials for Swedish patients with colorectal cancer undergoing elective surgery: A mixed method design. *Patient Education and Counseling*, 2014. 94(2): 202–209.

Solheim, O. et al., Incidence and causes of perioperative mortality after primary surgery for intracranial tumors: A national, population-based study. *Journal of Neurosurgery*, 2012. 116(4): 825–834.

Steinø, P., C.B. Jørgensen and J.K. Christoffersen, Psychiatric claims to the Danish Patient Insurance Association have low recognition percentages. *Danish Medical Journal*, 2013. 60(8).

Struwe, J. et al., Healthcare associated infections in university hospitals in Latvia, Lithuania and Sweden: A simple protocol for quality assessment. *Euro Surveillance*, 2006. 11(7): 167–71.

Suhonen, R., Individualized care in a Finnish Healthcare organization. *Journal of Clinical Nursing*, 2000. 9(2): 218–227.

Suhonen, R. et al., Impact of patient characteristics on orthopaedic and trauma patients' perceptions of individualised nursing care. *International Journal of Evidence-Based Healthcare*, 2010. 8(4): 259–267.

Suhonen, R. et al., Nurses' perceptions of individualized care: An international comparison. *Journal of Advanced Nursing*, 2011. 67(9): 1895–1907.

Sumanen, M. et al., Use of quality improvement methods in Finnish health centres in 1998 and 2003. *Scandinavian Journal of Primary Health Care*, 2008. 26(1): 12–16.

Söderberg, S., M. Olsson and L. Skär, A hidden kind of suffering: Female patient's complaints to Patient's Advisory Committee. *Scandinavian Journal of Caring Sciences*, 2012. 26(1): 144–150.

Thor, J. et al., Evolution and outcomes of a quality improvement program. *International Journal of Health Care Quality Assurance*, 2010. 23(3): 312–327.

Thorstad, M., I. Sie and B.M. Andersen, MRSA: A challenge to Norwegian nursing home personnel. *Interdisciplinary Perspectives on Infectious Diseases*, 2011. 2011.

Thue, G. et al., Quality assurance of laboratory work and clinical use of laboratory tests in general practice in Norway: A survey. *Scandinavian Journal of Primary Health Care*, 2011. 29(3): 171–175.

Tvedt, C. and G. Bukholm, Healthcare workers' self-reported effect of an interventional programme on knowledge and behaviour related to infection control. *Quality and Safety in Health Care*, 2010. 19(6).

Ukkola, A. et al., Patients' experiences and perceptions of living with coeliac disease – Implications for optimizing care. *Journal of Gastrointestinal and Liver Diseases*, 2012. 21(1): 17–22.

Unbeck, M., O. Muren and U. Lillkrona, Identification of adverse events at an orthopedics department in Sweden. *Acta Orthopaedica*, 2008. 79(3): 396–403.

Unbeck, M. et al., Healthcare processes must be improved to reduce the occurrence of orthopaedic adverse events. *Scandinavian Journal of Caring Sciences*, 2010. 24(4): 671–677.

Utzon, J., A.L. Petri and S. Christophersen, Analysis of quality data based on national clinical databases. *Ugeskrift for Laeger*, 2009. 171(38).

Vægter, K., R. Wahlström and K. Svärdsudd, General practitioners' awareness of their own drug prescribing profiles after postal feedback and outreach visits. *Upsala Journal of Medical Sciences*, 2012. 117(4): 439–444.

Vainiomaki, S. et al., The quality of electronic patient records in Finnish primary healthcare needs to be improved. *Scandinavian Journal of Primary Health Care*, 2008. 26(2): 117–122.

Vicente, V. et al., Elderly patients' participation in emergency medical services when offered an alternative care pathway. *International Journal of Qualitative Studies on Health and Well-Being*, 2013. 8(1).

Viskum, B., The hospital guidelines for medication reconciliation are not precis. *Ugeskrift For Laeger*, 2013. 175(1–2): 39–41.

Wallin, C.J. and J. Thor, SBAR – Modell för bättre kommunikation mellan vårdpersonal: Ineffektiv kommunikation bidrar till majoriteten av skador i vården. *Lakartidningen*, 2008. 105(26–27): 1922–1925.

Wallin, L. et al., Progress of unit based quality improvement: An evaluation of a support strategy. *Quality and Safety in Health Care*, 2002. 11(4): 308–314.

Wensing, M. et al., Tailored implementation of evidence-based practice for patients with chronic diseases. *PLoS ONE*, 2014. 9(7).

Wessel, M. et al., The tip of an iceberg? A cross-sectional study of the general public's experiences of reporting healthcare complaints in Stockholm, Sweden. *BMJ Open*, 2012. 2(1).

Wettermark, B., U. Bergman, and I. Krakau, Using aggregate data on dispensed drugs to evaluate the quality of prescribing in urban primary health care in Sweden. *Public Health*, 2006. 120(5): 451–461.

Wettermark, B., K. Nyman, and U. Bergman, Five years' experience of quality assurance and feedback with individual prescribing profiles at a primary healthcare centre in Stockholm, Sweden. *Quality in Primary Care*, 2004. 12(3): 225–234.

Wettermark, B., G. Tomson and U.L.F. Bergman, Kvalitetsindikatorer för läkemedel – Läget i Sverige idag. *Lakartidningen*, 2006. 103(46): 3607–3611.

Wettermark, B. et al., Financial incentives linked to self-assessment of prescribing patterns: A new approach for quality improvement of drug prescribing in primary care. *Quality in Primary Care*, 2009. 17(3): 179–189.

Wettermark, B. et al., Forecasting drug utilization and expenditure in a metropolitan health region. *BMC Health Services Research*, 2010. 10: 128.

WHO, Rational use of medicine. *WHO Drug Information*, 2007. 21(1): 27–32.

Wiig, S. et al., Investigating the use of patient involvement and patient experience in quality improvement in Norway: Rhetoric or reality? *BMC Health Services Research*, 2013. 13: 206.

Wilde-Larsson, B. and G. Larsson, Patients' views on quality of care and attitudes towards re-visiting providers. *International Journal of Health Care Quality Assurance*, 2009. 22(6): 600–611.

Wolf, A. et al., Impacts of patient characteristics on hospital care experience in 34,000 Swedish patients. *BMC Nursing*, 2012. 11: 8.

Zarchi, K. et al., Significant differences in nurses' knowledge of basic wound management – Implications for Treatment. *Acta Dermato-Venereologica*, 2014. 94(4): 403–407.

Index

G

H

I

K

M

N

O

P

Q

R

S

T

V

W

Milton Keynes UK
Ingram Content Group UK Ltd.
UKHW040107071024
449327UK00019B/887